FEAST AT YOUR FINGERTIPS

FEAST AT YOUR FINGERTIPS

The Ultimate Catalog of Fabulous Foods by Mail

by Stephen Manley

A Citadel Press Book
Published by Carol Publishing Group

A Citadel Press Book
Published by Carol Publishing Group
Citadel Press is a registered trademark of Carol Communications, Inc.
Editorial Offices: 600 Madison Avenue, New York, N.Y. 10022
Sales & Distribution Offices: 120 Enterprise Avenue, Secaucus, N.J. 07094
In Canada: Canadian Manda Group, P.O. Box 920, Station U, Toronto, Ontario M8Z 5P9

Queries regarding rights and permissions should be addressed to Carol Publishing Group, 600 Madison Avenue, New York, N.Y. 10022

Carol Publishing Group books are available at special discounts for bulk purchases, for sales promotions, fund raising, or educational purposes. Special editions can be created to specifications. For details, contact: Special Sales Department, Carol Publishing Group, 120 Enterprise Avenue, Secaucus, N.J. 07094

Manufactured in the United States of America

10   9   8   7   6   5   4   3   2   1

Photograph, page 125, courtesy of Michael Heintz, with maple wooden basket by Barret's Bottoms, Kearneyville, Wyo. Photograph, page 203, by Graphics Arts Specialist, Inc., Atlanta, Ga.

Library of Congress Cataloging-in-Publication Data

Manley, Stephen.
    Feast at your fingertips: the ultimate catalog of fabulous foods by mail / by Stephen Manley.
        p. cm.
    "A Citadel Press book."
    ISBN 0–8065–1379–9 (paper)
    1. Food—Catalogs. 2. Grocery trade—United States—Directories.
3. Mail-order business—United States—Directories. I. Title
TX354.5.M35 1992
641' .029'473—dc20   92–39022
                                                                              CIP

Contents

FEAST AT YOUR FINGERTIPS

Introduction

As you might imagine, compiling a book to represent a cross section of the mail-order food industry is a formidable task. Out of the fifteen hundred companies that I have assembled on lists ranging from gourmet chocolates to seafood, I have chosen over 150 businesses to feature in *Feast at Your Fingertips*.

The omissions do not in any way reflect poorly on the hundreds of other companies not included in this edition. An effort was made to offer the reader a smorgasbord of companies while trying not to overlap products (for example, three businesses that supply the same type of luscious edibles). So although you may be part of the growing trend of mail-order food shopping, and your source or sources are not listed, worry not. As long as they are providing you acceptable service and quality, that is all that is important.

For the thousands of famished folks who order food through the mail, *Feast* will be a convenient guide to purchasing anything from the best corn-fed beef to lobster pasta. The book's second aim is to help educate the reader. Between the company descriptions are informative pieces concerning nutrition, the origin of food types, and other writings to expand your knowledge of the food industry and different cuisines. Also included are references to further reading or sources for more complete information on the various subjects.

I hope you find many new and exciting culinary experiences in the wide assortment of companies providing mail-order foods. Or go ahead and just satisfy your craving for a double-deluxe fudge cheesecake. Whether you use the listings for gift giving, in thanks for a job well done, or to try to make grumpy Uncle Ed smile, you won't be disappointed with the companies in *Feast*. Bon Appetit!

The Mail-Order Industry

There is no doubt that ordering from a mail-order company is convenient. With the busy lives we all lead today—picking the kids up after soccer practice, working overtime, and trying to maintain friendships, all without forgetting a special birthday or anniversary—mail ordering easily fits our daily schedules.

Historically, catalog shopping is an American tradition predating the birth of this nation. Because Benjamin Franklin had so many other more important achievements, he has never been dubbed "the father of mail-order shopping," but he should be. His first published catalog, in 1744, offered nearly six hundred books through the mail. Franklin stressed the most important criterion of the mail-order business with the guarantee that "those persons who live remote, by sending orders and money to said B. Franklin, may depend on the same justice as if present."

Today mail-order companies must live up to the standard set almost two hundred and fifty years ago. If people are not pleased with their orders, they won't reorder, and the company will go out of business. To obtain complete satisfaction when ordering, note the tips later in this section to ensure that you receive what you want, when you want it, and that it arrives in good condition at the proper location. The old adage that "there are no dumb questions, except those you don't ask" certainly applies to mail-order shopping.

More than 90 million Americans rely on the convenience and diversity of products purchased by mail each year. With the advent of affordable, dependable overnight and second-day delivery services, the mail-order food industry has been able to blossom into a $1.2 billion market, with approximately 3,600 companies tempting your palate.

When you first consider ordering special cuts of meat or even peanuts by mail, you may appreciate the convenience but question whether the quality is substandard or just average. With my dozens of experiences, I can assure you that the items received not only equal those obtained by shopping at the best markets in a metropolitan area, but in many cases surpass those products in quality, taste, and freshness.

Like any industry, the good companies abound because they want to stay in business. But there are always a few businesses that are "rotten apples." If customer satisfaction is not received, these companies should be reported to the various arbitration groups listed in this section. Remember, mistakes can happen. Reputable companies will remedy the problem immediately. Generally, generous satisfaction-guaranteed policies offer money back or replacement because the company wants and needs your patronage. Some businesses rely on mail orders to a significant degree, and in some cases mail orders comprise almost all of their transactions. They truly value your patronage and want to ensure that each purchase is a pleasant experience.

As you shop through *Feast*, the variety of food products will amaze you. If you aren't a veteran catalog shopper, take time to read the tips that follow. Always make certain your order is clear to the company you are dealing with. This is imperative. Otherwise, you may receive what you purchased, but not what you wanted to order. Now wouldn't grumpy Uncle Ed be surprised if he received live lobsters instead of cooked lobster tails.

Tips for Shopping by Mail or Phone

Most mail-order food companies have similar guidelines for you to follow when placing an order by mail or phone. These are generally located on or near the order form that comes with the catalog. To ensure proper delivery and pricing, and to take advantage of specials or discounts, it is important to review this information before your purchase.

On the order form you will also find the company's guarantee, return policy, phone numbers for ordering or customer inquiries (many are 800 numbers), sizing charts, shipping information, seasonal limits, and sales tax information. The following are additional tips.

● When requesting a catalog, always ask for the latest edition. Many companies print only one catalog a year, so ask for an old one if it will be months before the new edition is available. September is when many businesses send their updated catalog.

● Check the company's guidelines before ordering. If there are any questions regarding their policies, call a customer service representative. They will be happy to clarify the details.

● Fill out the order form clearly. If you haven't ordered from the company before, it's wise to make a photocopy of your order form. Some customers keep copies of all their orders in case a problem arises or for tax purposes. Make a note if you are purchasing with a credit card and be sure you know the complete amount, including shipping charges, sales tax when applicable, and any additional costs.

● If you are calling the order in, decide what you want ahead of time and have the product numbers available to speed up the transaction. This includes having your credit card number handy.

● *Never* send cash. No reputable mail-order business will accept it.

● At holiday times many companies cannot supply 24- or 48-hour service for the same price, or at all. Often there are ordering deadlines. So for timely deliveries during these seasons, call early. Some companies even offer discounts if the order is placed before a certain date.

● There may be some restrictions on shipping perishables from state to state, particularly fruit. Check the company's policies carefully so that you will not be disappointed when trying to place your order.

● Ask about other services that may be offered. Often, gift wrapping and other amenities such as bibs, utensils, hats, gift cards, or a variation of the product being ordered can be arranged for an extra charge, but in most cases these are free for the asking.

● Check to see whether a signature is required for delivery. If not, be certain there is a safe place to leave the parcel so it will be protected from damage from rain, snow, theft, animals, etc.

● If you are participating in a "food-of-the-month" club, be certain you know when the shipments will arrive. Most businesses send a delivery confirmation card when it is a gift. Note the payment terms carefully and what exactly is being shipped. If there are dietary restrictions or other factors, such as delivering items to two different addresses (a summer and a winter address), make these instructions clear. Sometimes the company will substitute items of equal or greater value due to uncontrollable circumstances. If factors like diet are important, the company must be informed about them.

● Most companies have minimum orders. Splitting an order with a neighbor can be a solution if you are interested in only one or two items.

Always keep in mind that mail-order companies are at your service and want to provide the best possible shopping experience so you remain a satisfied customer. They cannot do this if they do not understand all your needs. You will find that no reasonable request is a problem, since they look forward to satisfying you and to obtaining future orders.

Protection When Shopping by Mail or Phone

When shopping by mail, the experience should not only be convenient, easy, and even fun, but should also be reliable.

All orders consumers place are protected by the Federal Trade Commission (FTC), the United States Postal Service, state regulations, as well as by guidelines set for the industry by the Direct Marketing Association (DMA).

The FTC regulations have two main provisions:

● The Mail Order Merchandise Rule, also called the Thirty-Day Rule, governs the shipping of merchandise, notification of any delay, the handling of refunds, and consumer's right to cancel. Merchandise must be shipped within thirty days of payment of the order, unless otherwise stated in the solicitation. The company must notify you if this is not possible (and give you a second shipping date), at which time you may cancel the order and receive a full refund. If there is no customer response to the company's notice, the order is still active. If the company cannot meet the second shipping date, you have the right to cancel for a full refund once again. No response from the customer on the second notice may, and probably will, cancel the order completely. Communicating with the purveyor to ensure delivery or cancellation of your order will eliminate any potential problems.

Please note: Orders placed over the phone and charged to a credit card are not covered by the Thirty-Day Rule, although the FTC is considering a revision that will extend such coverage.

● The Fair Credit Billing Act helps consumers resolve billing disputes with credit card companies on purchases made by phone or mail. Under this act, the credit card company is required to investigate the claim and settle any error. A copy of your order can be very important in these cases.

The United States Postal Inspection Service, the law-enforcement arm of the U.S. Postal Service, can help you resolve complaints when mail fraud is suspected. For more information contact your local post office.

The DMA, established in 1917, has a set of guidelines for mail-order companies. These determine the standards for the firms and protect the consumer. Many of the companies in *Feast* are members of the DMA, but do not rely on this alone as an indicator of the company's reputation. Being a wise consumer, certain of what you are ordering, is much more important.

If you have exhausted all the avenues for resolving a problem with a particular mail-order food company but haven't obtained satisfaction, the DMA offers the Mail Order Action Line (MOAL). They have been successful in obtaining satisfactory results in a high percentage of consumer complaints. To contact this *free* service, write to: Mail Order Action Line, Direct Marketing Association, 11 West 42 Street, P.O. Box 3861, New York, N.Y. 10163-3861. Be sure to include as much information as possible, copies of canceled checks or money-order receipts, along with a detailed description of the problems you experienced. The majority of the cases are settled within 30 days.

Name Removal Services

If you made mail-order purchases in the past, your name has most likely been traded or sold to other companies that supply similar products or services that might be of interest to you. This is not illegal. Yet if you would prefer not to receive unsolicited catalogs or phone calls, there are two free consumer name removal services available. Both are provided by the DMA.

● Mail Preference Service (MPS) removes your name from national advertising mailing lists. By submitting your name, you can reduce the amount of national advertising mail you receive. If the company petitioning you does not participate in MPS, they will have to be contacted directly, by you, and asked to remove your name from their lists.

● Telephone Preference Service (TPS) removes your home phone number from the many national telephone sales lists. When your name is submitted to TPS, it is placed in a file that is made available to firms on a quarterly basis. As a result, the calls may not cease immediately, but within a few months, the volume will decrease substantially. Local marketers generally do not participate in national name removal programs. Calling the company and politely requesting that your name be removed is your only option.

● As an additional service to customers, some companies have a policy of not selling their lists. They proudly announce this in their catalogs. These businesses should be applauded for their respect for your privacy and demonstrate clearly that their aim is to make your shopping experience a pleasant one.

To participate in MPS or TPS, write, MPS or TPS, Direct Marketing Association, 11 West 42nd Street, P.O. Box 3861, New York, N.Y. 10163-3861. Provide all the needed information for each service, including variations of spellings, address, apartment number, and zip code. Be sure to enclose your phone number and area code for the TPS solution to unsolicited telephone calls.

Using Feast at Your Fingertips

Using *Feast* is quite simple. Each chapter comprises a specific group of food purveyors. For example, chapter 1 contains companies which offer chocolates, nuts, and candies. Each business is highlighted by its regular and unique products so that you can get a feeling about the type of food items they supply.

Unless noted, the companies offer more products than are listed, in some cases, hundreds more. When you find an establishment that sounds interesting that you may want to purchase from, call for a free catalog to see their complete product line. If you are very interested in the foods offered and are likely to order in the future, ask for a free sample with the catalog. Many companies will be glad to tickle your palate, making it next to impossible for you not to make a purchase. Also, note the businesses that will give you a discount or free surprise if you mention you found their name in *Feast.* The code for these extras is part of each listing.

At the beginning of each chapter there is a brief description of the listings and other points of interest regarding the companies, products, and food type(s). Be aware that some of the food items overlap. A company that specializes in meat products, which is listed in chapter 7, may also supply a limited, but high-quality selection of seafood (normally in chapter 5) for their "Surf 'n Turf" package. For this reason, you should review each company to understand the extent of their product line. For instance, if you were looking for "chocolate-covered potato chips," you would not find them in the chocolates, nuts, and candies chapter. This happens to be a specialty item of a purveyor of meats.

Finally, when ordering mail-order sustenance items, never order directly from this book. *Feast* is compiled to aid you in finding a wide variety of food companies servicing a vast selection of food products. *Always* contact the company for the latest catalog and current price list, and be sure to follow ordering and payment instructions carefully.

All information in researching the companies has been checked as of press time, but it is—as the catalogs themselves will tell you—subject to change without notice.

Company Listing Guidelines

The following gives a sample description of a listing, the codes, and abbreviations used in *Feast.*

WOLFERMAN'S GOOD THINGS TO EAT

One Muffin Lane❶   800-999-0169❷ Exc KS
P.O. Box 15913   800-999-1910 KS
Lenexa, KS 66215   Discount: 10%❸

> If you use English muffins only as a utensil (to carry jelly, a tuna melt, or hamburger to your mouth), **Wolferman's** will change that habit. Featured are...❹

❶ Complete mailing address. Often the cooking and packing facilities will be at a different location. This can generally be found in the catalog.

❷ Phone and facsimile numbers. These are the numbers to use when you have questions, are placing an order, or are requesting a catalog. Unless you have set up an account with the company or have made arrangements to use the fax, contact the business for their guidelines on faxing material. Note: Never try to use fax numbers to reach a human voice; these phone lines are strictly for fax transmissions.

❸ Discount Code. Many of the companies in *Feast at Your Fingertips* offer a one- or more-time discount if you identify that you found their name in *Feast.* Some companies have committed to such discounts already and others will when you call, so always identify that you found the company in *Feast* to receive a special bonus or discount. Please note that they certainly are not required to offer anything.

❹ A brief description of the products offered by the company. The complete product line, in most cases, is much larger than the few example products noted. Companies also change the offerings without notice, so receiving the latest catalog is very helpful to ensure successful shopping.

Codes for Phone Abbreviations

Codes are located after the phone number:

CA—A state abbreviation signifies that this number should be used if you are calling from within this state.

Exc CA—Callers from states except the one indicated by the abbreviation use this number.

[414]—Customers within the area code given in brackets should use the number without the area code. For callers outside the area code, a second number is always given.

Collect—These companies invite you to call collect to save you the charge.

FAX—The fax number for orders and questions. Check with the company for guidelines before using these numbers.

Canada—The number to be used if calling from Canada. Not all companies have specific phone lines for these phone calls.

Ext. 7—Ask for extension number 7 when the phone is answered.

Corp—Corporate headquarters number, not to be used for orders or catalog requests.

Order—Number to place orders only.

I encourage you to go ahead now and enjoy the convenience and fun of ordering food through the mail. The following chapters are bound to have many items, from the everyday to the exotic, that will interest you.

Always be certain of what you are ordering and read the catalogs carefully. With the above simple guidelines, I am sure your experiences will be as pleasant as mine.

Chapter 1
Chocolates, Nuts and Candy

This section deals with the coffee-table favorites. The companies featured offer some of the highest-quality sweets and crunchy nibbles for gifts, thank-yous, and your private cravings. These palate pleasers should not be reserved for holidays, birthdays, or special occasions. The most delightful surprises arrive when that special someone is not expecting them. But heed this warning: All products taken in excess are heavenly hazards to your waistline.

ASTOR CHOCOLATE CORP.
48-25 Metropolitan Avenue
Glendale, NY 11385

718-386-7400
718-417-8823 FAX

Astor Chocolate doesn't sell chocolate candies. They produce the zenith in creativity for chocolate aficionados.

Chocolate shells, both dark and white, are this company's perfect creation. The shells are ready to eat or to stuff with anything you would consider appropriate for a sweet delight. Tempt your guests with a complementary after-dinner liqueur in the Astor Liqueur Cup, nestled in a gold foil cup for easy handling and eating. The one-half-ounce minidessert cup is an ideal size to fill with mousse or sherbet, or try this: Fill the shell with whipped cream, and float the shell in a cup of hot coffee. Ordinary coffee becomes a Viennese mocha delight.

If a dessert of edible elegance is called for on the menu, chocolate seashells heaped full of sherbet, ice cream, or mousse with fresh fruit will make dessert the memorable finale to most any meal.

With seven sizes and shapes in dark and white chocolate, there is no reason to ever have a simple dessert again, since Astor Chocolate is at your fingertips.

BATES BROS. NUT FARM   619-749-3333
15954 Woods Valley Road
Valley Center, CA 92082

Your gift-giving worries are solved with **Bates Bros. Nut Farm's** convenient gift packs.

At first, only walnuts were sold in the garage off the Bates's house, but it wasn't long before more products were added to the line, culminating in the many goodies carried today. Bates Bros. has everything from walnuts, pecans, cashews, macadamias, and pistachios to the freshest moist apricots, prunes, and dates to the most delicious variety of old-fashioned candies, preserves, honey, and much more to please your palate.

Choose from the assortment of gift-pack offerings or create your own with a number of products. If you ordered a couple of varieties of raw nuts, roasted nuts, dried fruit, candy, glaced fruit, and more, with a modicum of creativity and a few plates, you could design your own gifts for the folks at the office or family and friends. If time is of the essence, certainly a gift of English toffee, chocolate fudge, or country trail mix will be greatly appreciated.

 BOOMER'S OOGIES   413-445-7772
P.O. Box 2669
Pittsfield, MA 01202

What's an Oogie? You really don't know?!

Some folks think it's a rich, moist cookie, only better! Others think it's more like a dense and fudgy brownie, but better! Then there are those who bite into an Oogie and think it's like a luscious, sophisticated truffle, yet better!

Boomer's offers three flavors—Wonderful Walnut, Triple Chocolate, and Peppy Mint. The Bushes (the ones in the White House) received some recently. Why not you!

The combination of Ghirardelli chocolate and secret ingredients makes the perfect Oogie! And Boomer's is the only company that produces these delights!

Julia Child, Donald Trump, and many "common" folks have tried and enjoyed these cookie/fudgy/truffly treats, with no regrets.

DAVID ALAN CHOCOLATIER   800-428-2310
1700 N. Lebanon St.   317-637-3716 FAX
P.O. Box 588
Lebanon, IN 46052

European and American chocolate manufacturers have attempted to duplicate the texture and taste of freshly made Swiss chocolates, but they have not come close to **David Alan Chocolatier's** standards.

Swiss chocolates are deliciously different from most domestic chocolates. The creamy smooth centers, not overly sweet, are surrounded by exquisite real chocolate coverings. Each will melt in one's mouth with outstanding velvety smoothness to enrapture the taste of the chocolate lover. With this company, you can depend on the same quality.

Whether you want to present small gifts with two of their famous Swiss truffles, or you need a box containing over one hundred and sixty for an occasion, they are available in milk, dark, and assorted varieties. If the gift demands glazed pecans or macadamia nuts coated with fine chocolate, you can order a four- to fourteen-ounce size for yourself, then the gift.

ETHEL M CHOCOLATES
2 Catcus Garden Drive
P.O. Box 98505
Las Vegas, NV 89193

800-638-4356
702-451-8379 FAX

You may remember the "I Love Lucy" episode where Lucy frantically loads her pockets and mouth with bonbons to keep pace with a production line. This is not how **Ethel M Chocolates** operates; this company takes the business of making chocolates very seriously.

Established in 1978, the product line boasts over 50 varieties of prepackaged chocolates and the company is happy to personalize custom assortments for any occasion. Butter Creams, Chocolate Nuts, Toffees/Krisps, and Truffles are just a few of the confectionery options offered. Their own specialties—Watercolors Confections, featuring perennial favorites (nuts, butter creams, and caramels) enrobed in pastel or yogurt coatings—make unique gifts that will be cherished and memorable.

Most of the mouthwatering confections offered by Ethel M Chocolates use milk or dark chocolate. The products exclusive to the company's product line provide a very personal gift for the connoisseur or a special someone.

The History of Chocolate

*T*he culinary "time traveler" visiting Spain in the sixteenth and
seventhteenth centuries would quickly notice Spaniards' love of
drinking chocolate. This habit was adopted from the conquistadors,
who discovered the "secret" of chocolate from the Aztecs, and in the
Mayan culture, the cacao bean was worshiped as an idol and was
even used as a form of currency. Since the tropical cacao tree could
not be grown in Europe, cacao beans were soon perceived as a precious treasure by the Europeans, too. Imported from the West Indies,
an area controlled by the Spaniards, chocolate was a jealously
guarded commodity for over a hundred years.

By the 1700s, chocolate houses were as popular as coffee houses
throughout Europe. An important date in the chocolate chronicle is
1828, when a Dutch candy maker perfected a screw press that could
remove most of the cocoa butter from the bean. The result, cocoa powder, changed drinking chocolate forever and set the stage for the
development of the delectable chocolates we enjoy today. It was discovered that excess cocoa butter could be added back into the cocoa powder to make a smooth paste. In this form, sugar could easily be blended into the paste to produce "eating chocolate." By the nineteenth century, condensed milk and chocolate liquor (refined thick paste of
roasted cacao seeds that is heated to a liquid) were mixed in to make
milk chocolate.

A plate of lemon creams drenched in dark chocolate or almond
nut clusters coated in milk chocolate is the ultimate luxury. If you
indulge, are you throwing good health habits out the window? No!
Actually, most nutritionists agree that enjoying your favorite chocolate concoction on occasion provides a boost to your mood without
playing havoc with good health habits. Dr. Ann Grandjean, director
of the International Center for Sports Nutrition, in Omaha,
Nebraska, notes, "There's a lot more to eating than just the intake of
nutrients. We eat for social and psychological reasons, too. If you

avoid foods that taste good or that you love, eating will become a pretty dull routine. In fact, there really isn't such a thing as a 'good' food or a 'bad' food. With all foods, moderation is the key. You could eat too many carrots or drink too much milk, for that matter!
Because confectionery tastes so good to those of us with sweet tooths, we forget that a piece of candy has far fewer calories than other commonly eaten desserts, like pie."
So don't give up chocolate, even for your waistline. With all the work that has gone into perfecting them through the centuries, the least you can do is enjoy them prudently and sneak an extra when nobody is looking.

GOLDEN VALLEY NUT CO.   408-842-4893
170 Rucher Avenue
Gilroy, CA 95020

For a fine selection of top-quality nuts, contact the **Golden Valley Nut Co.**

This company specializes in California pistachios, but also carries a full line of other first-rate nuts. All of the products are dry-roasted without additives or preservatives and are available salted, unsalted, or raw. Try some of their signature nuts, pistachios. They are offered in natural (raw or salt-

ed), garlic, and jalapeño flavors. Either the one- or two-pound bag will make an excellent gift or perfect addition to any buffet.

To ensure freshness, these products are processed to order (they do not warehouse). If desired, they will even custom-roast and package the nuts to your needs. Try the almonds in natural, salsa ranchero, jalapeño, or garlic. Or make your own mixed nuts with macadamias, cashews, walnuts, pecans, and more.

 **HARDY'S FINEST** 800-537-3468
P.O. Box 1029 912-374-3322
Eastman, GA 31023

If you want special sweet treats and want them to be truly exclusive, **Hardy's Finest** will fill the order.

This pecan plantation has set a standard for not only nuts, but chocolates. They have an assortment of pecan log rolls that make excellent gifts and are a definite holiday treat. In bag candy, Milk Chocolate- and White Chocolate-covered Pecans are tasty, but mix them with the Milk and White Chocolate Pretzels and you will have an unbeatable after-dinner nibble or party favor.

Are you planning to send an extra-special box of candies? The Pecan Clusters and Pecan Divinity are far above the corner store's. Or choose the coconut patties, or the white or milk chocolate-covered pecan boxes to say something precious to someone special.

KARL BISSINGER FRENCH CONFECTIONS
3983 Gratiot Street   800-325-8881
St. Louis, MO 63110

If opening a gift box of chocolates is not an overwhelming experience—i.e., "Honey, this looks too good to eat!"—then you have not ordered from **Karl Bissinger's**.

When you order a box of French Truffles, Valentine Cookies, or French marshmallow chicks, each arrives packed handsomely in a gift box. The sweets are designed artistically with swirls of colored sugar, adorned with hearts, teddy bears, flowers, or one of many other designs appropriate to the holiday.

The "I Luv You" assortment is guaranteed to tickle any sweetheart. This box contains dark French creme mints with white and red hearts on top of each, and the phrase "I Luv You" spelled out with white monogrammed mints. But beware, this gift may be hazardous to your hips!

For a European finishing touch to a dinner party or special event, order a box of twenty-four sugar cubes, each topped with handmade motifs. There are four varieties: floral, Easter, Valentine, vegetable.

Among the dozens of other offerings by Bissinger French Confections is the Ultimate Pizza! This is a twelve-inch chocolate extravaganza complete with "crunchies, pecans, and coconut."

LAMMES CANDIES   800-252-1885
P.O. Box 1885
Austin, TX 78767

Everything from traditional favorites to some of the finest confections can be ordered from **Lammes Candies**, which began creating their sweets in 1878.

Try a flavor of the Southwest, the Texas Chewie. This delight makes an ideal gift or addition to your coffee table, as long as they last. Rich, buttery caramel surrounds the large Texas pecans to make a taste sensation. You can receive them in one-, two-, three- or five-pound packs. Or send the luscious divinity, with pecans folded in buttons of soft, white divinity. This gift is only produced around holidays. Orders will be accepted after September 1st.

What could better send your message than a box of kisses. The assorted kisses come in five flavorful choices, individually wrapped: mint, molasses, chocolate, cinnamon, and peanut butter taffy. Perfect any time of the year for entertaining friends or indulging yourself. Many other tasty treats are offered.

MISSOURI DANDY PANTRY   800-872-6879
212 Hammons Drive East   417-276-5121
Stockton, MO 65785   Discount: 10%

For a wide assortment of nuts and nutmeats for snacking or cooking, **Missouri Dandy Pantry** is ready and waiting for your order.

Nuts and more nuts. American black walnuts are one of their specialties and come in large, medium, and smaller sizes, already chopped for your favorite recipe. Pecans come in mammoth fancy roasted and salted and natural varieties, along with medium and small sizes. For snacking, make your own mixture with large cashews (salted or unsalted), California pistachios, macadamia nuts, whole almonds, sliced almonds, and more.

Combination burlap bags containing two or more one-pound bags of the flavorful, crunchy nutmeats are the perfect gift for the cook in the family or the "couch potato." Your choices can also arrive in decorative tins with one or more types of nuts. Don't overlook the nutty confections. Some are even made for sugar-restricted diets.

NUTS D'VINE   800-334-0492
P.O. Box 589   919-482-7857
Edenton, NC 27932

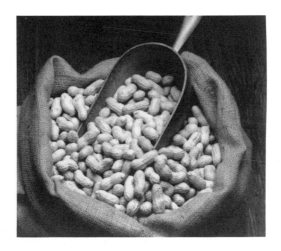

No party or Sunday afternoon ball game would be the same after you served **Nuts D'Vine.**

This company's peanuts and peanut products are sure to satisfy that peanut fanatic in your life. Raw Red Skins, a five-pound bag of them, will certainly put the peanut in the edible fantasy category. Or if three and a half pounds of Home-Style peanuts graced the coffee or buffet table, you can be sure they won't last. And when a special gift is in order, go for a bushel with the roasted or salted-in-shell five- to twenty-five pound versions that are certain to keep many mouths occupied for a long while.

One of the best traditional favorites is the old-fashioned peanut butter. Ground from freshly roasted peanuts, this healthy sandwich alternative can also be used in cooking. Unsalted extra-large peanuts are used in the Golden Peanut Brittle, and for stir-frying or your favorite recipes, there is no substitute for extra-large water-blanched peanuts.

Fannie May®

FANNIE MAY CANDIES   800-333-FMAY Order
1137 W. Jackson Blvd.   800-999-FMAY Help
P.O. Box 6939
Chicago, IL 60680

If you would like to experience some of the highest-quality candy, at a reasonable price, kitchen-fresh candies from **Fannie May** are your sweet tooth's choice.

Whether you are sending a thank-you gift, a little something to that special person, or satisfying your own personal cravings, one of the dozens of assorted chocolates will please the recipient. One-pound to five-pound selections are available. Try the assorted chocolates containing cream, nougats, toffees, and caramels with milk and dark chocolate coatings. Pixies with smooth, creamy caramel and crunchy pecans smothered with rich milk chocolate are perfect as an after-dinner treat.

Among the extensive collection of chocolates are buttercreams, fruit and nut, coconut, roasted nuts, and many other varieties. All come handsomely packaged in gift boxes or decorative tins. Gift certificates are also available.

Chocolate Trivia

● When was the chocolate bar introduced? *This snacking staple was developed in 1910 and gained popularity during World War II, when chocolate bars were issued as part of the infantry's D-rations.*

● What is the difference between milk and dark chocolate? *Milk chocolate is made with milk and between 10 and 15 percent chocolate liquor. The recipe for dark chocolate contains 35 to 50 percent chocolate liquor and no milk product other than butterfat.*

● Does chocolate "liquor" contain any alcohol? *No, it is simply the name given to the thick paste remaining after the roasted cacao seeds, or "nibs," are refined, pressed, and "conched" (heated to a liquid).*

● Is there an entree in which chocolate is unique? *Yes, one is Mole Poblano de Guajolote, Mexico's national dish, which contains turkey in a spicy dark chocolate sauce.*

● What is white chocolate made of? *White "chocolate" is a misnomer. It contains no chocolate liquor and is prepared from cocoa butter, sugar, milk solids, and vanilla.*

● Who eats the most chocolate? *The Swiss take the honors, averaging twenty-two pounds per-capita each year. England, Germany, and Belgium are close behind, while Americans eat fourteen pounds per person each year.*

MARY OF PUDDIN HILL   903-455-2651
4007 Interstate 30—Exit 95   903-455-4522 FAX
P.O. Box 241
Greenville, TX 75403

If you are in the mood for chewy, crunchy, crumbly chocolate treats, **Mary of Puddin Hill** is your place to shop.

This company offers a vast selection of chocolate covered caramels, toffees, pecans, Thingamajigs (caramels and pecans covered with chocolate), and more. All come in white and dark chocolate. Others include Ooooz 'n Ahhhz, which are Oreo* cookies covered with white or dark chocolate. They could add a new dimension to playing checkers.

One of the unique offerings is Little Puds. These are delicious individual servings of pecan fruitcake garnished with citron fruits. These are featured both along or in combination gift boxes. The mixed selections also may include delicious fudge balls, Chocolate Irresistibles, Fan-taste-ics, or truffles. The choice is yours, but it will be tough.

Mary of Puddin Hill's product line also has tortes, cakes, and cookies to dress up the dinner table.

*Oreo is a registered trademark of Nabisco, Inc.

MAUNA LOA MACADAMIA NUTS   800-832-9993
6523 North Galena Road   309-691-2632
P.O. Box 1772
Peoria, IL 61656

For a taste of Hawaii, the wide variety of exquisite paradise delights from **Mauna Loa Macadamia Nuts** will liven up your luau—even if you live in Connecticut.

Share the "aloha spirit" with a sampler of Hawaii's most delicious treasures. The eight-combination pack includes a complete assortment of macadamia nuts: three plain, butter candy–glazed, coconut candy–glazed, kona coffee–glazed, and honey roasted. Along with this combination comes chocolate-covered macadamia nuts, which are toted as Hawaii's "most treasured gift!"

Combination gift boxes with two to twelve cans will certainly say "thank-you" in a special way. Exotic-flavored hand-made truffles, for that extra-special favor, come in passion fruit, papaya, apricot, coconut, and macadamia nut. Perhaps a giant chocolate map of Hawaii is the perfect gift! Don't miss the Royal Kona Coffee selection, tantalizing dried fruits, jams, jellies, and desserts that Mauna Loa Macadamia Nuts exclusively offers.

MY SISTER'S CARAMELS   415-321-2582
1884 Bret Harte Street
Palo Alto, CA 94303

Whether you need a housewarming or hostess gift, or your personal caramel cravings have hit the roof, **My Sister's Caramels** will supply the elixir you need.

Their eight meltingly chewy flavors include: chocolate, coffee, praline (containing pecan pieces), vanilla walnut (with walnut pieces), plain vanilla, and peanut butter. The last two and newest flavorful treats have a soft spot of marshmallow in the center—vanilla mallo and chocolate mallo.

My Sister's Caramels received a rave review from Bon Appetit and was twice selected as having "the absolute best" caramels by both *Town & Country* and *Satisfaction Guaranteed.* One taste will show you why.

VIRGINIA DINER   800-868-NUTS
RT 460   804-899-2281 FAX
P.O. Box 1030
Wakefield, VA 23888

After sampling peanuts from **Virginia Diner**, you will drop the notion that the expensive grocery store versions are suitable for a Sunday football game or as nibbles at a dinner party.

This company's consistently crisp gourmet peanuts are available in salted, unsalted, roasted, or unroasted varieties. Honey-roasted almonds and butter-toasted peanuts can be ordered in one-pound bags or reusable decorative tins. (I use one of their tins to keep homemade cookies fresh.) Or, if you would like a large hug from a peanut lover, send them a peck or bushel of fresh peanuts in a decorative basket.

Peanut brittle, Crunches 'N' Clusters, and Peanut Duo are just some of the delights that will satisfy folks who enjoy peanuts, but also have a sweet tooth to satisfy.

The Crunches 'N' Clusters package includes peanuts glazed in a buttery coating of "crunches." The Clusters are a secret recipe of mouthwatering chunks blended with milk chocolate. Both are individually packaged in a decorative tin.

Many other baskets, combinations and foods are offered by Virginia Diner.

The Peanut

A baseball game or outing to the circus wouldn't be the same without somebody hollering, "Get your peanuts here! Peanuts!" And the unusually good taste of peanuts does not make you feel guilty if they're not on your diet. Who could ask for more?

But did you know that peanuts have a very unusual way of growing. Most folks know they grow underground, but not much more. Being a member of the legume, or pea and bean family, the peanut grows in pods with the peanut seeds inside. After the peanut plant flowers, pegs are formed when the blossoms wilt. Gravity pulls these pegs down into the ground, where the pods develop into peanuts. Each pod develops one or two seeds. These are then harvested, roasted, and then they are ready for eating at baseball games, circuses, and many other dining situations. As with most seeds (the peanut) and nuts, when used in cooking, they should be added at the end of the cooking time so they will not become soft.

Chapter 2

Cheese

When the discovery was made that fluid milk could be turned into cheese, that day was the dawn of one of the oldest of the prepared food industries. Most of the world's cheese is made from cow's milk. Yet in many parts of the world, goat's and sheep's milk cheese is a favorite at most meals. In fact, any milk-giving animal (mammalian animals), such as camels, horses, buffalos, reindeer, and others, provide milks that are turned into tasty varieties with distinctly delightful flavors.

Cheese is undoubtedly the "missing culinary link." I find a day doesn't pass without cheese complementing at least one meal. If only American cheese is in your refrigerator, and Swiss and cheddar appears only when guests are invited, your palate may rebel against the usual fare if it tastes Norwegian, Bulgarian, and African varieties.

If you have not tried quality aged cheddar or African Tuareg, the time has come to rediscover cheese's "missing link." The next few pages will help you in your quest.

BANDON FOODS, INC.   800-548-8961
P.O. Box 1668   503-347-2012 FAX
Bandon, OR 97411

During the 1920s, Bandon Cheese and Produce Company delivered cheese and butter by sternwheeler riverboats because there were few good roads. Today you can enjoy the delicious line of **Bandon Foods'** cheddar cheese and products, and you don't need to live on a river port.

Choose from the wide assortment of gift packs. Pride of Oregon includes an 8-ounce sharp, medium, garlic, and Baja in a gift box ideal for a housewarming or "thanks for the favor" gift. The Tradition, containing full cream sharp, extra sharp, smoked, jalapeño cheddar, and an eight-ounce beef stick, is a certain crowd pleaser.

For that special birthday or other occasion, send the Holiday Pack. With six helpings of cheese, the recipient will also enjoy "mini-mugs" of wild blackberry jam and cranberry jelly, all packaged handsomely in a wood logo box for arrival on Uncle Albert's birthday or at Joe's, who "came through."

Bandon Foods also offers many other gift-pack assortments of their special cheeses.

CALEF'S COUNTRY STORE   603-664-2231
Route 9, Box 57
Barrington, NH 03825

For traditional New England favorites **Calef's Country Store** is waiting to send a gift for you.

Their cheddar is famous and is shipped all over the world. Austin Calef describes the cheddar as "snappy. Makes a man sit up and take notice. All you have to do is to give a customer a sample on a cracker from the barrel and he's sold." This cheese can be on your table in one- and two-and-a-half pound waxed midget cheese blocks, three-lb. waxed midget cheese wheels, and an extra-sharp unwaxed cheddar in two-, three-, and five-lb. sizes. Smoky cheese is also available.

Other delights for gift giving include New Hampshire maple syrup from one-half pint to one gallon, Postcard Pack Maple Candy, bacon, jams, and jellies. If a special gift is needed, four gift packs combining the various country store items are certain to be a pleaser during the holidays or just to say thank you.

 CHEESE JUNCTION   800-253-3029
134 East Ridgewood Avenue   201-445-9211
Ridgewood, NJ 07450

If you always wanted your own cheese cellar with choices from around the world at your fingertips, **Cheese Junction** awaits your order.

This company brings more than one hundred varieties of cheese to you for both personal use and gift giving. All the cheeses are fresh, first quality, cut to order, and tailored to fit the needs of the individual or corporate buyer. Try some of the more than ten cheddars offered. Perhaps some German Limburger or Belgian Wynendale will make that European guest feel at home. Or to revive the memories of time spent in English pubs, English farmhouse cheddar with country bread and pickles might be your choice.

Many spreadable cheeses are available for that special hors d'oeuvres setting. Sharp cheddar with blue cheese, with horseradish and bacon, and with port wine are only a few of the cheddar varieties. The cream cheese collection includes garlic and herb, cheese roll swirled with red caviar, and garden vegetable. For those of you on sodium- and/or cholesterol-restricted diets, a number of selections are offered especially for you.

Gift box suggestions and also some of the world's finest coffees are part of the wonderful spectrum that Cheese Junction offers for you or the cheese-lover in your life.

 LAURA CHENEL'S CHEVRE 707-575-8888
1550 Ridgely Avenue
Santa Rosa, CA 95401

For meticulously handcrafted fresh and aged goat cheese, **Laura Chenel's Chevre** is the pioneer in this fast-growing segment of the cheese market.

In the fresh cheeses category they offer ten varieties. Some of these include Fromage Blanc, which is a smooth, light, tangy spread; Disk, which is coated with either herbs, black pepper, dill, or paprika; Pyramid, which is in a traditional French shape and is either plain or ash-coated. All the fresh cheeses have a mild flavor and are soft and spreadable in texture.

The seven selections of aged cheeses are "put up" to reach their "point" anywhere from just over a week to ten months. One of the choices is Tome. This is a firm cheese, in a wheel form, having a rich and mellow flavor. A second choice, Cabecou, is a nutty, densely textured hamburger-shaped cheese that is marinated with herbs in California extra-virgin olive oil. The aged cheeses have more complex, intriguing flavors and range from creamy to firm textures.

CROWLEY CHEESE 800-874-1239
Healdville, VT 05758 802-259-2340

From Vermont's oldest cheese factory, a National Historic Site, comes a wonderful assortment of regional cheeses. The bars and wheels from **Crowley Cheese** will remind you of authentic old New England stores, where a "wheel" was always stationed near the cash register and a taste would be handed to each customer.

Garlic, caraway, sage, dill, hot pepper, and smoked are among the appetizing selections, Flavored wheels (over two pounds) and taster's samplers (combinations of each flavored bar, plus a mild, medium, and sharp bar) will arrive in time for the holiday party or gathering with just a phone call.

Crowley Cheese has a wide assortment of gift packs, which range from the Holiday Plan (two deliveries) to a Deluxe Plan (four shipments). Special orders are always welcome!

About Cheese

I cannot imagine a world without cheese! A wise person once said, "Milk's leap to immortality was cheese." I will second that motion!

Whether a slice of blue cheese adorns a hamburger, grated Swiss is sprinkled on a warm chicken sandwich, or extra mozzarella is spread on lasagna, the culinary and workaday world would have a tremendous void without cheese.

A cheeseless world would make a ham and cheese sandwich— ham and bread. What would spaghetti and meatballs be without a healthy dose of Parmesan or Romano? Try imagining hors d'oeuvres of cheese and cracker—without cheese!

With a practically endless assortment of natural cheese (not cheese products) available today, three general categories emerge: hard, soft, and unripened. When you use cheese in recipes, you must follow cooking instructions carefully. Putting extra cheese in a souffle or quiche can turn your dinner into an edible mess! Yet extra cheese on pizza works wonderfully!

True hard cheeses are cellared. Like wine, aging brings the desired taste to a "point" at which the flavor stabilizes to a mild or sharp refinement. Refrigeration is not needed after aging hard cheeses. Yet all natural cheeses are constantly changing in taste. The robust or consistent taste is more noticeable and complementary in dry/hard varieties. Romano and Parmesan are good examples.

Soft cheese, mozzarella, for instance, is less stable than hard, aged varieties and should be eaten or sealed the day it is produced. Since a press and cheesecloth are not everyday equipment in American homes, the sealed grocery store varieties are more than adequate—and much more convenient than making homemade.

Unripened soft cheese, for example, cream and cottage cheeses, last only a few days if they are homemade. Store-bought varieties stand up better due to pasturation, but expiration dates must be observed carefully as well as how the store handles the cheese. The containers should never be left unrefrigerated. Spoilage is liable to occur earlier than the expiration date in these cases.

Exploring the world cheese is fun. If you have children, American cheese will probably be the staple in your home, but that should not hinder you from trying the tasty variations from around the world.

GETHSEMANI FARMS
Trappist, KY 40051

502-549-3117
502-549-8281 FAX

What was craved by Lyndon Johnson, mentioned in *People*, and served to the Pope? Do fruitcake haters have an option? The answers and delicious edibles come from only one source, **Gethsemani Farms.**

Trappist Cheese is the response to the first question. The unmatchable mild and aged cheeses come in three-pound wheels and single or gift packs. Or enjoy the trio of quarter wheels, including all mild, all aged and mild, and aged and smoky. Other options include half wheels and for the health-conscious, lower salt and lite Swiss cheese offerings.

When ordering cheese, you would be skipping two treats without Trappist Fruitcake and smooth Bourbon Fudge. Those who have passed fruitcakes back and forth among family members for years as a joke will have a knife and plate out when they receive a Trappist Fruitcake. For that fudge lover in your family, the Bourbon Fudge will make a mighty tasty gift.

Feast at Your Fingertips   **39**

MAYTAG DAIRY FARMS   800-247-2458
Box 806   515-792-1133
Newton, IA 50208

If the highest-quality Midwestern cheese is a must for your table, **Maytag Dairy Farms** is your source. With fifty years' experience in cheese making, their Holstein-Friesian herd has been the top award winner in North America for over twenty years. This excellence results in exquisite cheeses.

Try the sharp natural white cheddar on a cracker or with a slice of pie and you will realize how deprived your taste buds have been. The creamy mellowness of Maytag blue cheese does not just happen! Eight days pass from the time the cheese maker first pours the milk into the warming vat until he places the wheels in the ripening "caves" for six months! Note: This is twice as long as for ordinary blue cheese.

Middlefield® Cheese

MIDDLEFIELD CHEESE
Box 757
Middlefield, OH 44062

800-32 SWISS
216-632-5228 OH

For top-quality gifts of cheese for every occasion, **Middlefield Cheese** gives a splendid variety of solutions.

Bulk cheeses from this company in blocks, horns, or wheels make the perfect gift or addition to a party or buffet table. Try the mild Wisconsin muenster, rich and mild colby, or the famous Middlefield Baby Swiss. If you enjoy the delicate taste of yogurt, you will truly appreciate the mild and creamy yogurt cheese.

Or for a perfect thank you or congratulations send one of the firm's many baskets and packs. The Party Starter Basket comes in a wicker basket with three-quarter pounds each of Middlefield Swiss, colby longhorn, New York cheddar, brick, and a seven-ounce wheel of old-fashioned Gouda cheese. The twelve-ounce Ohio hickory-smoked summer sausage and other surprises make this gift perfect for any occasion. There are many other combinations to choose from.

THE PLYMOUTH CHEESE CORP.
Box 1
Plymouth, VT 05056

802-672-3650

For a taste of Old New England cheeses and other treats, **The Plymouth Cheese** company is at your service.

Among the regional products are the three- and five-pound wheels of true, old-fashioned Vermont Granular Curd Cheese. The wheels are delivered in four varieties: regular, sage, pimento, or caraway. You can also accent the parcel with 100 percent pure grade A medium amber maple syrup, in one-pint to one-gallon sizes.

Sample a Cheese and Pickle Packet, including watermelon pickle, mustard pickle, pepper relish, corn relish, and three one-pound wedges of regular, sage, and caraway Plymouth cheese. Or experience a taste of New England in one of their many other distinct selections.

TILLAMOOK CHEESE   800-542-7290
P.O. Box 313   503-842-4481
Tillamook, OR 97141

Over the years, **Tillamook Cheese** has remained true to three fundamental principles: contented cows, quality products, and satisfied customers. This has to be an unbeatable combination.

The Baby Load, which brought Tillamook its fame, is just two pounds of all-natural and richly delicious cheddar in medium, sharp, and low-sodium varieties. Five-pounders, perfect for sandwich-size slices, hors d'oeuvres, or sliced with apple pie are available in medium and sharp cheddar. Or try the sinfully rich spreads in soups, baked potatoes, tossed with pasta, or even on crackers. The flavors include wood-smoked cheddar, jalapeño cheddar, cheddar pecan, and sharp cheddar.

For a gift or sensational nibbles on a hike or ski weekend, the Oregon County Picnic Basket combines over twenty items for nibbling by friends or as a perfect picnic for two. A salmon fillet, spreads, ten-ounce bars of Monterey Jack, sharp and medium cheddars, Country Smoker Summer Sausage, and Haus Barthe Mustard are just a hint of the taste treats you will receive.

SAY! CHEESE   800-368-2437
P.O. Box 308
350 Main Street
Somerset, WI 54025

For the best of factory fresh cheeses from Wisconsin, **Say! Cheese** will deliver and demonstrate the large difference between "factory fresh" and store-bought varieties.

With over 30 types of cheeses in one- to five-pound blocks, this company is perfect for supplying especially fresh products for every occasion from small gatherings to large banquets. Each cheese is shipped with directions on how to store the cheese, how long it will keep, some background on the cheese, and suggested uses.

A full line of gift boxes for any occasion is also available along with bulk cheese bargains.

Discovery of Cheese

A ccording to a legend passed down through the ages, about 2000 B.C. an Arab merchant named Kanama was traveling through the desert. Milk, being the chief beverage, was carried in a flask. On this particular day, Kanama decided to bypass his noonday meal and continue his arduous journey until nightfall. Being tired and thirsty, Kanama placed his flask to his lips, expecting a refreshing drink of milk, but a trickle of thin, watery liquid flowed forth. Angered and determined to find the reason, he cut the flask, which was a dried sheep's stomach, and noticed a soft white mass. The process that altered the milk involved a natural substance known today as rennin (a digestive enzyme found in the stomachs of animals). Kanama tasted the curd and found it to be delightfully flavorsome. Thus, according to the legend, Kanama became the first *person to eat cheese. Upon reaching his destination, he related his experience to his friends and in this unusual manner, cheese was discovered.*

All in all, the legend probably contains a modicum of truth. Kanama is not around, so the experience is hard to verify. But there is no doubt, due to the worldwide production of cheese since ancient times, that on all continents there were "Kanamas" who accidentally discovered that milk solidifies when warmed under certain conditions. Since the process takes about ten quarts of milk to make a pound of cheese, the yield was very nutritious and easily transported. To this day, cheese making remains one of the oldest of the prepared food industries.

SEGUINS HOUSE OF CHEESE   715-735-9091
W 1968 Highway 41
Marinette, WI 54143

If you are looking for a wide variety of cheese and cheese-and-sausage gift packs, you should not miss the ample selection offered by **Seguins House of Cheese.**

A quick gift of two, five, or eight cheeses (the Family Pack) will be delivered across the country and bring a taste of Wisconsin to relatives in Maine. Many of the packages can be tailored to the needs of the individual. Cheddar cheese shaped as a barn, a cow, or the state of Wisconsin can be suitably arranged with mustards and strawberry candies. Other additions to the gift packs are summer sausages and/or hot wheel sausage!

Sequins can also ship "wheels" of cheddar in three-, six-, or 3 one-pound sizes. All can be ordered in mild, medium, or sharp flavors. Baby Swiss is available in a "round" style or block. For those who have restricted diets or are more health conscious, order the Lacey-Swiss for a low-salt, low-cholesterol, low-fat version. Sausage and beef sticks can also be part of your delivery.

An assortment of cheeses in a crock round off the selection. When ordering, don't forget Seguins House of Cheese's bulk bargains!

VELLA CHEESE COMPANY   800-848-0505
315 Second Street East   707-938-4307 FAX
P.O. Box 191
Sonoma, CA 95476

For over sixty years **Vella Cheese Company** cheese selections have been an accompaniment to the finest foods, given as gifts, and even enjoyed on Mount Everest.

While much has changed in sixty years, the yeast cultures, the personal hands-on techniques, even the old curing room are the same. Choose from the more than thirty items in wheel or block form, two to twelve pounds. A variety of Monterey Jack styles, cheddars, and Oregon blue cheeses can liven your buffet table or make a perfect thank you.

When ordering, you should not overlook the true masterpiece, Dry Monterey Jack. An International Gold Medal Winner in 1988, this cheese is excellent in casseroles, toasted cheese sandwiches, omelettes, and soufflés. It is also wonderful shredded on soups, pastas, or tacos.

Chapter 3

Desserts and Baked Goods

Mother Nature certainly has a sense of humor. If it were up to her, we would reach for a celery stick during dinner to complement the roast beef and mashed potatoes. For dessert, she would be glad to see us eat a couple of lettuce leaves and stay thin.

In reality, we reach for a muffin or two, while keeping a slice of bread close by to sop up the gravy. And as our sweet tooth screams for a third piece of cheesecake, we show great restraint by only nibbling on the edges as it goes back into the refrigerator. Why couldn't Mother Nature let muffins and cheesecake be like lettuce? In digesting lettuce, your body must use more calories than it gains.

So as you go through this section—have the last laugh on Mother Nature. Enjoy the sinfully delicious desserts and don't miss the wonderful assortment of baked goods. Anyone for a chocolate chip muffin?

Amber and Company

AMBER AND COMPANY   800-635-4994
P.O. Box 2204   816-373-3999
Independence, MO 64055

Is there someone you should send a long-stem box of cookies to? If so, **Amber and Company** can help you fill the order.

In your order, you will send a decorative long-stem roses box with your choice of a dozen delicious jumbo cookies. All will be wonderfully arranged to surprise your special someone with cookies that will tell your personal message. Telecookies come in four flavors: chocolate chip, peanut butter, and oatmeal raisin, the newest being gourmet chocolate chip with English walnuts and pecans.

The box of one-dozen jumbo Telecookies is the perfect choice when you want to send a unique thanks, enjoyed meeting you, or Mother's Day gift that will be remembered with fondness. The cookies are also available in decorative, reusable tins. Don't overlook the newest additions of the cake offerings: pound, fruit, banana nut, and chocolate turtle.

BEN & JERRY'S   802-244-6957 Ext. 423
Route 100   Discount: Free Gift
P.O. Box 240
Waterbury, VT 05676

Product Mission—To make, distribute, and sell the finest-quality all-natural ice cream and related products in a wide variety of innovative flavors made from Vermont dairy products! **Ben & Jerry's** has succeeded!

We all know at least one ice cream fanatic, if not dozens. When those enthusiasts have a bad day, week, or perhaps they fixed your flat tire, free, send them a couple of pints of Ben & Jerry's Vanilla Chocolate Chunk, or Heath Bar Crunch. If it was raining when the tire change occurred, some Rainforest Crunch is probably in order.

Each of the sixteen flavors available in pints will change your conception of what truly superior ice cream is like. Pints of your favorite flavors can be sent through the mail in specially designed packages that keep the treats frozen.

Ben & Jerry's also donates part of the proceeds to any number of charities. So when you give a gift of Vermont's finest all-natural ice cream, you may be helping in world peace efforts or the ongoing homeless problems.

BETTE'S DINER PRODUCTS   510-601-6980
4260 Hollis #120   510-601-6989 FAX
Emeryville, CA 94608

If, in your travels, you enjoyed the delicious meals at Bette's Ocean View Diner, in San Francisco, you can savor the same treats in Virginia or elsewhere from **Bette's Diner Products.**

Italy may have its espresso, Paris its cafés...but Berkeley, California, has discovered something even better...an honest cup of Good Strong Coffee. The best-kept secret of Berkeley is Bette's Blend with five dark-roasted beans; you can also order regular or decaf. One cup is certain to bring back memories of San Francisco.

Bette's scone and pancake mixes have satisfied thousands of customers and should be part of your kitchen supplies. Choose the Raisin or Wild Cranberry Scone Mix for unforgettable baked delights, or the combination of pancake mixes, including buckwheat, golden cornmeal, and oatmeal. A combination would make a perfect gift for a newly transferred San Franciscoan.

 BRUMWELL'S FLOUR MILL   319-622-3455
South Amana, IA 52033

From its small beginning in 1936, **Brumwell's Flour Mill** has grown steadily and today can supply your home and gift-giving needs.

This company has a wide variety of products milled from yellow, white, and blue corn, wheat, rye, oats, buckwheat, barley, and soybeans. Your choices are extensive. They also make various pancake, muffin, biscuit, and granola mixes and have a *natural* line, including basic syrups, pure sorghum, and apple butter. All are perfect gifts for folks who bake.

Recipes are included with many of the products. Try or send some Pancake and Waffle Mix; you'll never again be satisfied without it. Or choose the Dark Rye Flour or Soy Flour. Perhaps the Seven-Grain Bread Mix for old-styled bread or the Biscuit Mix will suit your needs All will make special gifts and certainly will make cooking easier.

CAFE BEAUJOLAIS   800-332-3446 Exc. CA
Box 730   707-964-0292 CA
Mendocino, CA 95460

Along the northern California coast, where majestic redwoods and clean air are a given, **Cafe Beaujolais** prevails, turning out country-kitchen tastes that can not be matched.

Straight from the bakery comes Panforte Di Mendocino in four rich and chewy flavors: almond, hazelnut, walnut, and macadamia. Or try a dried fruit fruitcake. This delight was created after someone commented that nobody can make a decent fruitcake. The delicious dessert doesn't contain anything green or red that you want to pick out and absolutely nothing candied. Dark, spicy, and filled with dried fruit, they come in two varieties: original and chocolate. These bakery items are perfect for a hostess, housewarming, and other gift-giving needs.

Send for a bag of Spicy Gingersnaps or Homemade Cashew Granola to bring a smile to someone who needs a lift. There is nothing better than a Gingersnap with a glass of milk for a snack or a bowl of Cashew Granola with milk to start the morning. Maybe the sweeter, marvelous creation Chocolate Almond Toffee is the perfect gift.

If you are in the area, you can also visit the Cafe Beaujolais restaurant for an unforgettable meal.

C'EST CROISSANT, INC.   800-633-2767 Exc. CA
22138 S. Vermont Avenue #F   800-833-2767 CA
Torrance, CA 90502

If you are, or know, a croissant aficionado **C'est Croissant, Inc.** is your place to shop.

They offer eight convenient packs with ten to twelve buttery delights in each. The C'est Croissant Classic includes six rich, flaky all-butter and six plump, puffy French almond croissants, with three jars of fruit preserves and a woven French country tea towel. Or for a celebration of sophisticated tastes, the Love Dozen contains six sweet raspberry-filled and six sweet-tooth satisfying chocolate croissants.

COLLIN STREET BAKERY   800-248-3366
401 W. Seventh Street   903-872-8112
Corsicana, TX 75110   903-872-6879 FAX

If you resent the cutting remarks you hear about fruitcake, try the paragon of the business, **Collin Street Bakery.** You will have the last laugh!

Since 1896, the Deluxe Fruitcake offered by Collin Street Bakery has been made true to the Old World recipe born in Wiesbaden, Germany. Today the cakes are shipped to all fifty states and overseas to 195 countries. This alone demonstrates the true quality and character of this afternoon tea staple, or holiday dessert.

The three sizes range from a small two-pound package to a party-size five-pound cake. Each Deluxe Fruitcake is delivered in a decorative holiday tin, which includes a description of the colorful history of the Collin Street Bakery.

COUNTRY EPICURE   612-938-1949
Vie De France Corporation   612-938-0443 FAX
11562 Encore Circle
Minnetonka, MI 55343

Receive an upscale cake for a special dessert or congratulations from **Country Epicure**, and you will partake of the finest.

The Midnight Layer Cake defines the perfect chocolate dessert. Four layers of a dense, moist, and fudgy cake are filled with a superbly smooth chocolate whipped cream. The Carrot Layer Cake is simple pleasure at its best. The stunning gift or dessert treat is fragrant and lightly spiced; this flavorful carrot cake has a crunchy walnut texture that is unbelievably gratifying and is topped with a cream cheese topping spiked with a hit of rum.

The many other delicious offerings include Mississippi Fudge Cake, Black and White Espresso Cake, Velvet Cake, Raspberry Creme, and more. If your guests are amazed at your dessert, send them one on their birthday. For cheesecake fans, the White Chocolate or New York–style will surely make a hit.

DESSERTS DIRECT, INC.   800-44-CAKES
78 Mt. Vernon Street
Ridgefield Park, NJ 07660

Your next dessert is ready and waiting for your call at **Desserts Direct, Inc.**

For a happy birthday greeting, congratulations, or because you don't feel like baking, these desserts are ideal for the most discriminating palate. Select one of the Classics: Bombe Au Chocolat or perhaps the English Charlotte Royale, which is a light Genoise cake rolled with raspberry purée and filled with a delicate raspberry and vanilla bavarian cream. The final touch is the cake's apricot glaze. Even if you claim not to have a sweet tooth, you must know someone who would enjoy this delight.

Maybe one of the Gourmet Cakes will end the perfect meal with panache. Some of the choices are: Black Forest Cake, banana walnut torte, carrot, or German chocolate cake. All can be described with the word delectable and will make perfect additions to your dining experience or a fine gift that the recipient will not soon forget.

DESSERT OF THE MONTH CLUB   800-423-3091 Exc. CA
16633 Ventura Boulevard   818-501-6363 CA
Encino, CA 91436

Your sweet tooth (or teeth!) will "stand" at attention with this company's products. Chocolate Chip Crunch Bars, Orange Sunburst Cake, and Zesty Lemon Squares are only a hint of the after-dinner delights offered by the **Dessert of the Month Club.**

A "Party in a Box" might be your thing! The Birthday Box includes: chocolate chip fudge cake, a large bow to place on the cake, balloons, party hats, and noisemakers. "Custom-made for the occasion," is Dessert of the Month Club's motto, and they will tailor a sweet-tooth fantasy for anniversaries, Valentine's Day, Thanksgiving, Christmas, Easter, and most other celebrations.

The sweet delights can be ordered last minute, or you can establish a plan to have delectable treat arrive monthly. The Dessert of the Month Club is deliciously adaptable to the needs of corporations, small groups, and holiday gift giving—even your private cravings!

Some Common Baking Substitutes

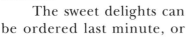

Even the most stocked cupboards run out of something when you are in the middle of a recipe. Substitutions are generally a matter of looking for a similar food, or even the same one in a dried state. The results will be adequate if not unnoticeable in most cases.

● Baking Mix, Commercial: Combine two cups of all-purpose flour, 1 tablespoon of baking power, 1 tablespoon of sugar, and 1 tablespoon of salt. Cut in 1/3 cup of solid vegetable shortening to a course grain consistency.

● Buttermilk: Combine 2 tablespoons of vinegar or lemon juice with 1 cup of skim or 1 percent milk and let stand for five minutes. Or replace buttermilk with an equal amount of low-fat plain yogurt.

● Cream Cheese: Blend 1 cup low-fat cottage cheese and 1/4 cup margarine. This is equal to 1 cup of cream cheese.

● Eggs: To eliminate a yolk, use two egg whites. Most recipes will not be affected.

● Herbs: Note whether the recipe says dried or fresh herbs. If using dried, and the recipe calls for fresh, cut the amount by one third. Dried seasonings will lose their potency after a year, so add to taste, but be careful—you can't remove them once they're added.

● Milk: When making cakes or muffins, fruit juice can often be used in place of milk. If the juice is acidic, add 1/2 teaspoon of baking soda. If you only have skim milk, and need a whole milk equivalent, add two tablespoons of melted butter or margarine to 1 cup of skim milk.

● Sour Cream: Purée 1 cup of cottage cheese or ricotta with approximately 2 tablespoons of yogurt or buttermilk to desired consistency. Or whip 1/2 cup chilled evaporated milk with 1 tablespoon of white vinegar to make 1 cup of sour cream.

GRANDMA'S FRUIT CAKE   800-228-4030
Box 457
Beatrice, NE 68310

The **Grandma's Fruit Cake** is claimed to be "The World's Finest Fruit Cake."

Try the traditional fruitcake. Every slice is filled with delectable raisins, fresh walnuts and pecans, cherries—all surrounded by the unmistakable taste of fine rum. They come in ring cakes or loaves ready to serve or send as a gift. The secret recipe uses only the finest ingredients: rich ripe cherries, tender Malaysian pineapple, quality walnuts, pecans, raisins, and ingredients that bring the fruitcake alive.

Try a light fruitcake for an ideal after-dinner dessert— Mrs. Craver's Light and Dark Fruit Cakes, or Grandma's original Amaretto Cake. Each Amaretto Cake comes in a European-designed collectible tin—perfect for weddings, birthdays, or any holiday.

GRANNY BOOZER'S CHEESECAKE
P.O. Box 885   800-848-7057
Anniston, AL 36202

Start a tradition or have a very special treat with one of the delectable cheesecakes from **Granny Boozer's Cheesecake**.

In the foothills of Alabama's Cheaha Mountain, you will find this company's special kitchen. Only the freshest dairy products, natural flavorings, and premium liqueurs make the moist, rich cheesecakes come alive.

Try the Heavenly Honey, Absolutely Amaretto, or Cosmopolitan Chocolate as a priceless luncheon or afternoon tea dessert. Or the elegant Glorious Grand Marnier or luscious Magnificent Margarita with a lime "zing" may be the perfect topping to a perfect dinner. For newcomers to Granny's try the Granny's Cheesecake Sampler, which arrives with eight flavors that will surely tempt and then delight the most discerning palate.

GRAY'S GRIST MILL   508-636-6075
P.O. Box 422
Adamsville, RI 02801

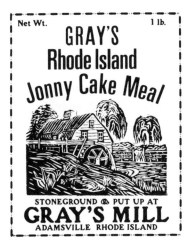

Since 1878 the grindstones at **Gray's Grist Mill** have been turning and producing meal, flour, and mixes for the New England area. Now you can enjoy the traditional tastes through the mail.

Although their selection is limited to six varieties at this time, each of them will add a new dimension and taste to your favorite recipes. Pancake and Waffle and Brown Bread and Muffin Mix are perfect to adapt to breakfast recipes. For breads, they offer whole wheat and rye flour. For frying try the Jonny Cake Meal or the new addition to their line, Fish and Clam Fry Mix.

All the items come with suggested recipes and uses. All of Gray's Grist Mill products are 100 percent plain whole food. Note that their selections are expanding, and they offer custom grinding also, so give them a ring today for the most recent catalog.

H & H BAGEL   212-595-8000
2239 Broadway   800-692-2435
New York, NY 10024

There are claims around the country that a truly good bagel is like finding the "hole" left in a donut! With **H & H Bagel** at your fingertips, you will find the "hole!"

Maybe plain or poppy will work for Monday! Onion or Sesame Tuesday! On Wednesday perhaps pumpernickel or whole wheat! Salt or sourdough are the favorite on Thursday! Cinnamon-Raisin or garlic will keep the office, friends, and family under control on Friday!

When the weekend arrives, try some mini bagels. Perfect for hors d'oeuvres or finger sandwiches at a luncheon. Then enjoy the leftovers from the week. H & H Bagels rated second in the *Boston Herald*'s Taste-Off, despite the fact that they were purchased five days before, then frozen and thawed twice before presentation.

Imagine the quality when shipped fresh, overnight!

HAYDEL'S BAKERY   800-442-1342
4037 Jefferson Highway   504-837-0190
New Orleans, LA 70121

Nationally famous **Haydel's Bakery** is waiting to custom-ship a fresh, delicious Mardi Gras King Cake to your favorite out-of-town reveler.

For over thirty years this company has been treating discriminating palates to the finest pastries, cakes, and Mardi

Gras King Cakes which consist of sweet dough braided with cinnamon and sugar, topped with white fondant icing, and purple, green and gold sugars. The package also includes the annual *Haydel's New Orleans Mardi Gras Magazine*, a King Cake history scroll, Carnival beads and doubloons, and while the mold lasts, a handcrafted porcelain King Cake doll. This is more than just a special cake and will add fun to any party.

Choose from the many other New Orleans baked goods. Praline cheesecake or olde-fashioned Russian cake will make a delicious ending to any meal or a perfect gift for a special occasion. Or try the treat that is ideal with breakfast, brunch, or dinner—Cajun Kringle. This flaky, buttery pastry has an unique pecan filling and is topped with caramel icing and decorated with pecans. The Legend of Cajun Kringle is included.

HEARTY MIX CO.   908-382-3010
1231 Madison Hill Road   Discount: 10%
Rahway, NJ 07065

For a large variety of baking mixes, providing good nutrition and good taste, review the offerings of the **Hearty Mix Co.** You will not be disappointed.

There are eight categories of mixes offered. The following are samples from each. Yeast Bread Mixes: buttermilk, whole wheat, italian, and rye. Quick Bread and Pancake Mixes: bran-wheat germ, corn meal, and Boston brown. Cake Doughnut Mixes: buttermilk and whole wheat buttermilk. Also included is information on calories and sodium per serving for each mixture.

For people on salt-restricted diets there are many options. Wheatless Bread and Muffin Mixes, Wheatens, Fake Rye, and Wheatens Baking Mix contain 13 percent sweet cream buttermilk powder. All are salt-free! Low-Sodium Mixes expand your options with varieties such as whole wheat buttermilk bread, coconut almond macaroon, date-nut bar, and many more.

Hearty Mix offers many other extra-hearty/naturally fortified prepared mixes to suit any baking requirement. For good-tasting quick-and-easy breads, muffins, and croissants add a mix or two to your shopping list soon.

LU CLINTON'S SUGAR SPOON   800-228-0052
1885 West A Street   308-532-1484
North Platte, NE 69101

Fantastic! Fabulous! Cheesecake delivered presliced, all decorated and ready to serve. Each and every **Sugar Spoon** cheesecake is fussed over by Lu, for you.

Only the finest ingredients are used...velvet smooth cream cheese...country fresh eggs...fine liqueurs...rich chocolates...toasted pecans and almonds. Try some of the favorites: strawberry, peaches 'n' cream, or double chocolate amaretto. For an afternoon brunch, offer cinnamon-caramel-pecan, classic plain, or Bailey's Irish Cream.

Some of the new flavors offered are strawberry margarita, white chocolate-raspberry truffle, and turtle, which combines buttery caramel, smooth dark chocolate, and toasted pecans intertwined with white creme de cacao to make this a pièce de résistance.

MAID OF SCANDINAVIA   800-328-6722
3244 Raleigh Avenue   612-927-6215 FAX
Minneapolis, MN 55416

Whether your specialty is making cookies, candies, traditional ethnic specialties, or decorating cakes and petit fours, you will be able to find many of the ingredients and equipment through the **Maid of Scandinavia** catalog.

For the candy maker in your house, fondant, meringue powder, confectioner's glaze, candy dipping centers, and more are offered. The dipping centers, for instance, are available in ten flavors ranging from whipped chocolate to lemon redi-Center. If ready-made candies are needed, choose from the wide assortment of Pretty Candies for entertaining and gifts or Sweets For Your Parties. These assortments of crowd pleasers are sure to be a perfect addition to any occasion.

For baking and desserts, Maid of Scandinavia offers many mixes and icings for household or even commercial uses. Doughnut, Commercial Cake, and Pound Cake mixes are perfect for the special treat of fresh doughnuts at home or if you have to make a wedding cake. All the supplies for large cakes that you can imagine are available.

MISS GRACE LEMON CAKE CO.   800-367-2253
422 N. Canon Drive   818-995-1976
Beverly Hills, CA 90210

The company's original lemon cake is loved because it is made with juice from fresh lemons. **Miss Grace Lemon Cake Co.** offers even more options to your dessert choices.

Order the Chocolate Fudge Cake: warning—for chocolate lovers only. Crisp California walnuts and real chocolate morsels enhance this irresistible choice. The French Liqueur Cake is delicately blended with fine imported orange liqueur, orange juice and peel, and finished with a drizzle of Swiss chocolate on top. It will soften up anyone, or could make a perfect congratulations gift.

Elegant gift baskets contain a sampling of the Original Lemon Cake, Chocolate Fudge Gracelet (mini version of the cake), the sensational California Orange Gracelet, plus Miniature Macadamia Muffins. Don't miss the delicious cookies in tins or the monthly offerings. Send a delicious treat on a four-, six-, or twelve-month schedule as a constant reminder that you care or as a seductive offering on your behalf.

MORGAN'S MILLS 207-785-4900
RD 2, Box 115
Union, ME 04862

Morgan's Mills specializes in the freshest organic flour products, stone-ground by water power.

Try one of the muffin mixes: Blueberry, Maple Corn, and the moist and zesty Orange Bran Muffin Mix. Griddle-Waffle Mixes include Buttermilk, Buckwheat, and Rice, Corn and Oat Griddles which are wheat and dairy free. Try these and other offerings when baking the morning muffins, or send a gift to the special baker; also offered are many other griffles, mixes, meals, and flour variations.

For other gift needs, choose from the Mothers Mountain Mustard or Nilsdotter Bean Soup Mixes selection. From Nomad Apiaries, you have ten or more varieties of honey: raspberry, blueberry, autumn blossom, and antique cork are just a few. Many specialty jams and other edibles are also available.

MRS. PEABODY'S COOKIES 313-761-2447
715 North University Discount: Free Samples
Ann Arbor, MI 48104

If you are one of those rare individuals who doesn't like to nibble on cookies, even occasionally for breakfast, don't read about **Mrs. Peabody's Cookies** scrumptious selection.

In the Chocolatier Collection, your chocolate fantasies will be satisfied. This combination of semi-sweet chocolate chip, white chocolate, and fudgy chocolate chip cookies is certain to fulfill your sweet tooth's desires. The Hearty Collection includes semi-sweet chocolate with walnuts, semi-sweet chocolate with peanut butter, and oatmeal-raisin cookies. If you cannot make up your mind on these and the other delightful cookies offered, try the Variety Sampler, which contains 4 each of the 6 scrumptious flavors.

Except for the Variety Sampler, all the "collections" come in reusable gift tins containing one or two dozen cookies. Other cookies are available, providing an assortment for every taste.

OREGON MOUNTAIN PRODUCTS　　800-237-6047
0333 S.W. Nebraska　　　　　　　　　503-246-8875
Portland, OR 97201　　　　　　　　　Discount: 10%

Devastatingly delicious would be an insult to **Oregon Mountain Products**' rum cakes. They have set an unsurpassable standard with their original (vanilla rum) and chocolate (chocolate rum) varieties.

Delivered in four-pound, two-pound or 5.75-ounce sizes, these cakes are perfect for an afternoon tea or a fancy dessert. The small "Souvenir" size comes in a four pack that is ideal for individual portions at a dinner party or for a starving college student! The larger sizes are tailored for larger gatherings or more portions.

 SAN ANTONIO RIVER MILL　　800-627-6455
P.O. Box 18627
San Antonio, TX 78218

For down-home Southern baked goods, from biscuits to pancakes, and more, **San Antonio River Mill** is your source for some of the best.

Their vast selection of baked goods mixes will challenge the baker in your kitchen. The Blue Chip Breakfast is a combination of two breakfast favorites, Blueberry Pancake Mix and Sweet Cream Waffle Mix with a container of blueberry syrup. The Muffin Maven's Assortment will challenge you on which tender muffin to make first: chocolate chip, honey bran, apple and cinnamon, or sweet corn. All are hard to resist when topped with Guajillo honey (included). If biscuits are needed for dinner with gravy, or a strawberry shortcake dessert, don't ignore the variety of biscuit mixes.

Other offerings include: Traditional Southern Cornbread Mix, Jalapeño Cornbread Mix, Apple and Cinnamon Muffin Mix, Cinnamon and Raisin Biscuit Mix, and a large selection of preserves, jellies, syrups, and many other condiments. Southern smoked meats, fish, and game are also available.

SANTA FE COOKIE CO.
110 W. San Francisco Street
Santa Fe, NM 87501

800-243-0353 Exc. NM
505-983-7707 NM

Sending the goodness of homebaked cookies has never been easier than with **Santa Fe Cookie Co.**

The cookies from this company are fresh. The tasty morsels are packaged for delivery in decorative tins within 12 hours of baking. Choose from Shortbread with Piñon nuts, Coconut Macaroon, Peanut Butter Chocolate Chip, Chocolate Mocha

---

Mint, and more. As a gift for a starving college student or a thank-you for housesitting, these cookies will be certain to make the recipient smile.

You can send a single tin for a special occasion or join their exclusive Cookie-of-the-Month Club and treat yourself and friends to cookies throughout the year. Each month or every other month, one variety will be sent to satisfy that someone, and leave a lasting and continued reminder of your gratitude, love, or flair for the unusual.

Mail Food Clubs

" *I*t's delightful; it's delicious; it's delovely, " sings the character in the hit Broadway musical Anything Goes! *The character was not referring to mail-order desserts—but probably should have been.*

Many nationally known food suppliers are making our pants tighter, delightfully! These companies offer "food-by-mail" clubs. You can sign up for a monthly or bi-monthly treat or a number of special occasions or holiday packages. And if you just need a onetime gift to keep "grumpy Grandpa" smiling, they will be glad to oblige.

The companies listed in Feast are reasonably priced, provide excellent quality, and their shipments arrive promptly. Special orders are not usually a problem, given reasonable lead-times.

The perfect desserts for "grumpy Grandpa" or the new client you "landed" are just a phone call away to the listed companies. Individual food items or even combinations can be designed according to your needs. Special preparation, greeting cards, or if a personal/embarrassing touch is necessary, most "food-of-the-month" clubs will include your "literature," pictures, or other materials for the special package.

Whether the event is a milestone, for instance, an employee's retirement, or your mechanic finally found the problem that makes your car stop dead, or perhaps Uncle Ed is up to the usual, a specially prepared package can be devised by these companies. Or, if Uncle Ed needs a regular reminder, create a plan with the many variations available.

If you try a food-by-mail club, you may be pleasantly surprised, but please see Chapter 1, the section titled Tips for Shopping by Mail or Phone, for more information and guidelines for mail-order food clubs.

TURF CHEESECAKE CO.   914-668-9176
158 S. 12th Avenue   212-654-1622
Mt. Vernon, NY 10550

Turf…"New York's Rolls-Royce cheesecake maker." *Esquire,* July 1982.

Yes! Their cheesecakes are that good! Eight of the most sinfully delectable flavors are only a phone call away. Plain,

---

chocolate chip, or amaretto are perfect for ending an afternoon lunch or with morning coffee. Brandy Alexander, topped with mocha cookie crunch or chocolate containing a special blend of sweet chocolates are the ultimate "fixes" for chocoholics!

For the holiday season, or just because of it…pumpkin and cranberry will add to a festive occasion or afternoon treat. For an authentic Old Country delight order the Black Forest. This is a black cherry cheesecake topped with chocolate cake crumbs. Your private sweet cravings will be satisfied by this nirvana-like experience!

Many other delicious surprises are available from Turf Cheesecake, Co. Old-fashioned-style tarts in raspberry, apricot, nutty pecan, or chocolate brownie can be on your table this week with just a phone call. They can be served hot or cold as a perfect finish to any meal.

WILLIAM GREENBERG JR. DESSERTS, INC.
1377 Third Avenue   800-255-8278 Exc. NY
New York, NY 10021   212-861-1340 NY

If you are an international celebrity, or just like me, **William Greenberg Jr. Desserts** is bound to be a pleaser.

Your craving or gift-giving needs may lead you to "The Brownie" of New York. Or the three classic butter cookies are perfect for any table: the Thumbprint filled with seedless raspberry jam, crushed pecan shortbread rolled in sugar, and the Orange Diamond covered with slivered almonds and cinnamon sugar. Perhaps the refined cheese straws—elegant puff pastry twists flavored with finely grated Vermont cheddar in plain, sesame, and caraway flavor will satisfy you.

If dessert demands cakes, pastries, or muffins, your order can be completed quickly. The pound cake is ideal for entertaining and comes in loaves and bundt style—plain,

raisin, chocolate pecan, chocolate chip, and others are available. Try some of the seven flavors of four traditional shapes of Danish pastries. Or a nontraditional cake-like muffin which comes in blueberry, cranapple, corn, and bran with raisins.

WOLFERMAN'S, GOOD THINGS TO EAT
One Muffin Lane   800-999-0169 Exc. KS
P.O. Box 15913   800-999-1910 KS
Lenexa, KS 66215

If you use English muffins only as a utensil (to carry jelly, a tuna melt, or a hamburger to your mouth), **Wolferman's** will change that notion. Featured are fourteen flavors of world-famous English muffins. Also offered are holiday gift tins, mealtime snack kits, and scrumptious cheese and sausage assortments. All will tickle your palate and upgrade your opinion of English muffins.

Grocery store selections do not compare to Wolferman's wide variety—garlic, jalapeño, corn, sourdough, cheddar cheese, blueberry, and onion pumpernickel are some of your choices. The full lineup includes seven additional mouthwatering varieties, all of which are thicker than "standard" English muffins. Their bite-size Mini English Muffins are a unique flavorful base for hors d'oeuvres, tasty dipping pieces for a fondue, or will add zest to any cracker and cheese plate.

Don't ignore the special occasion tins or gift baskets! Design your own from the diverse selection, or choose from Wolferman's fifteen fun-filled options!

The "Eggs Benedict" tin includes 10 packages of Original English Muffins, two pounds of delicious Canadian bacon, and Hollandaise Sauce Mix with simple preparation instructions. Half-orders are offered!

The "Snack Basket" includes a picnic wicker basket to carry the Sourdough and Original English Muffins, sherry mustard, all-beef summer sausage, and a sampling of sharp sage-and-smoked cheddar cheese.

Chapter 4

Regional and Ethnic

We all like different regional and ethnic cuisines. Unusual tastes wake up our palates and induce our internal chefs to pep up an ordinary dinner entry. Unfortunately, sometimes the "doctoring" leads to peculiar results. But if you start with the right ingredients, or better yet, have someone who knows what they are doing prepare the dish, the results can make for a memorable meal.

This chapter will introduce you to many companies that offer prepared meals and hard-to-find ingredients and seasonings. With these references, you will no longer have to watch your dinner guests swallow your newest creation with a remorseful smile. They may actually ask for a second helping. Just don't tell them you sprinkled chocolate-covered ants on the dessert and you're home-free.

ALASKA SAUSAGE & SEAFOOD CO.   907-562-3636
2914 Arctic Boulevard
Anchorage, AK 99503

If you have been looking for something tasty but different to serve at your next dinner party, take a look at **Alaska Sausage & Seafood Co.**'s offerings.

Established in 1963, they have become well known for processed game meats. Moose, caribou, and reindeer are trimmed, boned, aged, and prepared into steaks, roasts, and sausage. Think of the look on your guests faces when you ask them, "Would you like caribou or moose sausage with your eggs?" All these tastes are ready for the grill, oven, or pan and include cooking instructions.

In the early seventies the seafood division of the company was started. They now offer both smoked and fresh canned salmon in three varieties: red, king, and silver. Also available are One Side Smoked Salmon (kippered), Smoked Red Salmon Fillet (18 ounces), and a Smoked Fish Sampler, including kippered salmon, one can of smoked salmon, smoked halibut, and smoked salmon strips.

ALASKA WILD BERRY PRODUCTS   907-235-8858
Box 374
Holmer, AK 99603

Would you like a taste of Alaska in your kitchen?

Alaska Wild Berry Products supplies a wide variety of native products. Spread a layer of strawberry, blueberry, or wild raspberry jam on your morning toast or muffin. Try some Sourdough Bread with Lowbush Cranberry sauce for lunch. Or, start the evening's hors d'oeuvres with Alaskan salmon and end the meal with an Alaska Wild Berry Chocolate or two!

Gift packs in handsome wooden boxes can contain your choice of products from Alaska—whether Polar Sticks (reindeer meat) or cranberry tea, or perhaps some smoked cheddar, is on the menu.

BACINO'S 708-655-1982
75 E. Wacker Drive
Chicago, IL 60614

Most of us have called the local pizza palace for a quick supper or accompaniment with a movie. But if you plan ahead, you can enjoy award-winning Chicago deep-dish style pizza from **Bacino's**.

Not only will you receive one of the country's finest pizzas, but the tasty delight is healthy. Started as a pioneering concept in a town where pizza is king, the Heart Healthy stuffed spinach pizza is the only stuffed pizza to merit the American Heart Association of Chicago's imprimatur as a healthy entrée.

Their Heart Healthy stuffed spinach pizza blends two part-skim milk cheeses packed between a light pastry-like crust, vegetable or olive oil, imported sauces, and bunches of fresh spinach to create an unforgettable no-salt-added meal. This heart-shaped pizza sets new quality standards that you just won't find in other pizzas.

BEST OF KANSAS   316-685-5566
5426 E. Central
Wichita, KS 67208

If you are short on time, but long on consideration and taste, **Best of Kansas** has a gift box or basket to fit your needs.

For that outdoor barbecuer, send the Buffalo Bill BBQ Sampler, including some of the nation's top-rated BBQ sauces—all from Kansas. Mom's BBQ Sauce comes in three flavors: garlic and onion, hot and spicy, and hickory smoke. In addition, cooks can order Golden Mill Sorghum, K.C. Rib Doctor Seasoning, and Hayward's Pit Bar B-Q Sauce. This

gift is certain to add many hours of cooking and eating enjoyment.

If you are sending a gift to someone whose idea of hot food means beads of sweat on their brow, the Wild Bill Hickok Hot n' Spicy is a must. This includes Kansas Wheat House Wheat Nubs (Cajun), Art's Hot Tater Chips, Toma Picante Mix (HOT), Land of Ahs Jalapeño Popcorn Seasoning, Rabbit Creek Old-Fashioned Kansas Chili, and much more.

The Best of Kansas offers many other gift packs and main meals to suit most any of your gift giving needs.

BEWLEY IRISH IMPORTS   215-696-2682
1130 Greenhill Road   215-344-7618 FAX
West Chester, PA 19380

For a wide variety of Irish specialty foods, try **Bewley Irish Imports** and have a touch of Ireland at home. Bewley's imported Irish foods include: cookies/crackers, teas (loose/bagged), Irish whiskey marmalades and preserves, mustards, shortbread, fruitcakes, and more. A large assortment of cookies/crackers is offered and would complement the twelve Irish teas and six coffees.

Or try the mustard selections, which includes Irish whiskey, Guiness Stout, horseradish, and honey. They are bound to add a taste treat to a sandwich or sauce.

For that after-dinner or morning luxury, order from their wide selections of marmalades. Offerings include Vintage Irish Whiskey or Irish Mist and eight varieties of Laird's Irish preserves, jellies, and marmalades.

BURRITO EXPRESS   800-553-8388
1597 E. Washington Boulevard   818-798-0844
Pasadena, CA 91104

If the closest emporium for burritos is like mine, an hour away and not worth the side effects, try the authentic Southwestern treats from **Burrito Express.**

After receiving numerous requests from travelers and former Pasadena residents, these restaurateurs developed the Burrito Six-Pack. For six mouthwatering burritos you can request beef, chicken, or pork flavors prepared fresh, then flash-frozen and shipped by overnight express. They come in mild and spicy versions and arrive frozen, ready for lunch or dinner.

These large, plump burritos are not the usual offerings that you can eat three or four at a sitting. Each is a tasty meal by itself.

BYRON PLANTATION   800-241-7013
P.O. Box 1599   404-333-0383 FAX
Byron, GA 31008

Delicious plantation tastes from the heart of the Deep South can be yours from **Byron Plantation.**

A plantation gift to friends or your own table could include the Southern Sampler. The gift box includes vidalia onion relish, peach butter, and Scruppernong (grape) jelly. Maybe the Plantation Special with toasted and salted pecans, chocolate pecan caramel clusters, spiced pecans, and chocolate-covered pecans will please the "nutty" recipient better. The generous assortment of Georgia pecans prepared in four delicious ways should not miss.

Choose from the pails of peanut treats. How about an authentic peck of peanuts? Or peanut butter and roasted and salted peanuts, which are certainly always a pleaser. If your sweet tooth needs a treat, try the Teejays, which are candy-coated peanut sweets, or the Crickle in a Tin offering a fix for the peanut- and pecan-brittle lover in your life. Maybe it's you.

CHAPITA'S, INC.   415-591-6559
P.O. Box 6840
San Carlos, CA 94070

Chapita's, Inc. presents: "A Taste of the Southwest," and you are invited.

The president of this company recalls how people were sized up in New Mexico: "They based your fortitude there on how hot you like your chili. It's not good unless it makes your nose run and your brow sweat." It is evident that the sauces, salsa, marinades, and gazpachos are serious—and only for those who are not expecting traditional Mexican but pine for the Pueblo Indian influence in their food.

Try one of the three or five packs for home use or as a gift to a friend who has never experienced real "hot stuff." The Delicioso Five Pack includes: Fajita Marinade, Chili de Abuela, Salsa Aztec, and Gazpacho and Chili Verde. If you prefer to start with medium heat, try the Chile Rojo, for an enchilada sauce, and

Salsa Nopales. But don't skip the Fajita Marinade with pineapple and lime juices and spices; it is ideal with poultry, fish, and meat.

DAKIN FARM   802-425-3971
Route 7
Ferrisburg, VT 05456

For the homesick traveler or a "touch of Vermont" at your next gathering, **Dakin Farm** can supply you with the best of Vermont to make the occasion special.

For a breakfast treat, the Green Mountain Breakfast includes a full pound of cob-smoked bacon, a pint of pure Vermont maple syrup, and a two-pound sack of Dakin Farm buttermilk pancake mix, which will please any hearty appetite. Smoked Canadian bacon and cob-smoked sausages are among the other delicious breakfast meats that can adorn your table.

Smoked hams for lunch or dinner might be more your preference. Order them with the bone in, boneless, or ready-to-eat for the buffet table or sandwiches at lunch. Try a smoked turkey breast sandwich, or enjoy the whole bird. You may not be royalty, but eat like it with ringneck pheasant for an entrée or appetizer. The perfect dessert with these offerings might be Vermont maple syrup drizzled over vanilla ice cream.

Dakin Farm offers gift-of-the-month clubs with many appetizing items in different plans. Contact them today for some Vermont cheddar cheese or whipped cheddar spreads to complement your above choice.

EARLY'S HONEY STAND   615-486-2230
P.O. Box 908   615-486-1121 FAX
Spring Hill, TN 37174   Discount: 10%

In 1925 **Early's** began curing and smoking Tennessee Hill Country pure pork sausages, hams, and bacon. To this day the old-time methods are not compromised when the meats are hung and slowly smoked over hickory embers.

Try a sausage breakfast package. The selection includes: three-pound bag of buttermilk biscuit mix, a two-pound poke of hickory-smoked pork sausage, a twelve-ounce jar of

---

clover honey, and a ten-ounce jar of blackberry jam. Fresh
link sausages can be substituted for the smoked sausage during certain months. Perhaps a thick-sliced or center-cut
bacon breakfast including griddle cake mix could turn the
ordinary "Weekend Wheaties" meal into an experience!

Early's also offers smoked lunch and dinner enchant-
ments with their whole and half country ham, country ham
squares for sandwiches, whole turkeys, and turkey breasts!
Try a combination package that includes Tennessee cheddar
cheese, barbecue sauce, and a smorgasbord of Early's
smoked meats.

When ordering don't miss out on the Chow-Chow
blends, the Sweet and Tart Pickled Combo, or the many
other Tennessee country cupboard selections.

FURTADO'S   800-845-4800
North End Provision Co., Inc.
544 North Underwood Street
Fall River, MA 02720

The true art of "old world" chourico and linguica lives on at
Furtado's.

To enjoy the finest, leanest Portuguese sausage made,
you will have to place an order. The chourico is an exquisite
surprise atop pizza. After one tasting, you will never order
regular sausage again. Or if you just choose chourico for
lunch, a quick delicious sandwich is just minutes away.
Whether you fry or boil chourico, a bit of mustard, pickle,
and cheese on Italian bread with the meat makes an extra-
special lunch, meal, or snack on poker night.

Linguica with pasta, linguica with eggs in the morning—the options are unlimited. Both the chourico and linguica are available in patties, franks, cocktail franks, and
sausages. The sausages come in medium and hot varieties.

Morcelas (blood pudding), a traditional favorite, is offered on a seasonal basis.

G. B. RATTO & COMPANY　　800-325-3483 order
821 Washington Street　　　　800-228-3515 Service
Oakland, CA 94607　　　　　　510-836-2250 FAX

If you have been looking for an international grocery selling everything from sauces to seeds, **G. B. Ratto & Company** features imported foodstuffs not found in your local markets.

With its smorgasbord of products mixing Old World cuisines with the best in domestic cooking, Ratto is an international retail food store at your fingertips. Try the Asian, Spanish, or British specialties. Maybe a charming present is needed for a "French garlic lover." No problem. Anchovy paste from Spain or sauerkraut and red cabbage from Germany might put the unmatched touch to a meal or bring old memories to the surface.

Try the tasty pesto sauce. Or for salads to meatloaf, order pearled barley, buckwheat groats, or bulghur wheat. The nuts and dried fruit section offers delights from Australia to Turkey. Or simply feast on ginger snappers from San Francisco.

GAZIN'S　　　　　　　　502-482-0302
2910 Toulouse Street　　　　504-827-5319 FAX
P.O. Box 19221
New Orleans, LA 70179

If finger-lickin', rib-stickin', great-tastin' Cajun/Creole food is on your menu, let **Gazin's** satisfy your desires.

The tastes of Louisiana, whether in sauces, condiments, or full meals, can adorn your table without moving to Cajun

Country. For that extra-special appetizer or mainstay spice, use the Cajun Power Garlic Sauce, Pecan Butter Sauce, or Mardi Gras Dip. Add some lump crabmeat for an authentic touch of Louisiana.

Perhaps to warm up January the gumbo, scampi, and pasta would be suitable. Red beans and rice will get you past Groundhog's Day. In March, take advantage of the monthly special meal: okra gumbo, Cajun-Creole etouffée (a combination of shrimp, crawfish tails, or chicken in a roux-based brown sauce), French bread, and Mississippi mud chocolate dessert. Each month a specialty is featured in the Meal-of-the-Month packages.

Gazin's offers numerous other spices, sauces, picnic packs, and other edibles that will take you to Cajun Country after moving to Kansas, or introduce the New England palate to the wonderful tastes of the South. (Note the sidebar on Cajun cooking, page 86).

K-PAUL'S
824 Distributor Row
New Orleans, LA 70123

800-4KPAULS Exc. LA
504-731-3590 LA
Discount: Sample

For the truly delicious tastes of Cajun cooking, go to the source, **K-Paul's** in New Orleans. Since most people cannot sample a traditional Cajun meal at world-famous chef Paul Prudhomme's restaurant in Louisiana, Paul ships Cajun delights directly from his kitchen to your table. The three "Mail-A-Meals" include Jambalaya, Etouffée, and Shrimp Creole.

Jambalaya is a hearty, wonderfully spicy rice dish filled with Cajun meats (andouille and tasso), and either chicken or seafood. Etouffée is a combination of a seasoned brown roux-based sauce filled with shrimp, crawfish tails, or chicken—served over rice. Shrimp Creole is a blend of fresh gulf shrimp and Creole tomatoes with Magic Seasoning Blends to create a delicious red sauce, all served over rice.

In addition to the meal selections, K-Paul's offers smoked turkeys, pecan praline, sweet-potato pecan pie, and a wide assortment of Magic Seasoning Blends for the experimenting cook. Cajun meats such a andouille and pork tasso are also available.

Cajun Cuisine

*M*any cultural influences can be found in Cajun cuisine. This cooking genre can be traced to five groups of settlers and the natives of what is now Louisiana.

The French influence is the most prominent. When the Cajuns were routed from their Acadian (Nova Scotia) homeland, they brought their French culinary arts with them. The Spanish introduced red pepper, via Mexico, and probably the eating of rice. Garlic came from the Italians, and German settlers contributed mustard. Africans brought okra; West Indians sent over peppers and allspice. And the native Choctaw Indians discovered and used filé (dried leaves of the sassafras tree) long before the Acadian Coast was settled.

This melting pot of cooking is what today gives us the spicy, zesty tastes of Cajun fare.

KONRIKO   800-551-3245
Company Store
307 Ann Street
New Iberia, LA 70560

South Louisianians say that a day without rice is a day on which you haven't eaten. If you "haven't eaten" lately, let **Konriko** end your "fast."

Their offerings include Original Brown, Wild Pecan, and 1912-Style Bran & Wild Rice. Or, Cajun Rice Pilaf, Artichoke Rice Mix, or jambalaya mix with rice may be more suitable. All the rice can be bought by the case, and many of the styles in two-, five-, or ten-pound burlap bags. Some products are boxed.

Other traditional Cajun treats are also available. Cajun golden honey, King's ten-bean soup with seasonings, Creole seasonings for seafood, and a variety of sauces, syrups, and other edibles will introduce you to the true tastes of Cajun cooking. Combination gift packs can also be ordered.

LA ROMAGNOLA   800-843-8359
2720 N. Forsyth Road   407-679-1907
Winter Park, FL 32792   407-657-0168 FAX

La Romagnola products are designed for the professional chef. If that doesn't scare you away, keep reading.

This company produces twenty-five varieties of noodles and filled items as well as a unique line of sauces. The filled pasta has a ratio between filling and dough—no less than 40 percent filling by weight. Included are black and white lobster ravioli, spinach ravioli, cheese ravioli, triangoli with mushrooms, potato gnocchi, and more. The linguine-style noodles come in tomato, spinach, black, and egg noodle flavors, to name a few.

Using their Sauce Plus products, you are able to reduce preparation time when making fine sauces without sacrificing quality or taste. Sauce Plus doesn't contain any meat and can therefore be used in vegetarian dishes. Sauce Plus is available in five flavors and several sizes. These flavors are porcini mushroom (enhances delicate flavors of chicken and veal), red pimento (perfect on seafood or poached fish), basis, tomato, and pesto sauce.

All of La Romagnola Sauce Plus offerings are shipped frozen and are easily defrosted.

MADE IN WASHINGTON   800-338-9903
1530 Post Alley   206-728-9044 (206)
Seattle, WA 98101   206-728-2298 FAX

To explore new tastes, or cure a homesick traveler, **Made in Washington** is sure to do the trick.

Serve Port Chatham Smoked Salmon or Cougar Gold Cheese (only available in November and December), or perhaps farm-fresh Aplets or Cotlets to have a sample from the Pacific northwest. Enjoy the selection of shelled hazel nuts, preserves, or red delicious applies handpacked and shipped immediately.

Try the best of the Northwest gift packs, including a state sampler with delicious foods from the four corners of the state. Savor the Pancake Perfect for a Sunday breakfast or send this sampler complete with topping to friends and relatives. Or for the angler, a chocolate Fighting Trout or Northwest Salmon to celebrate, or ease, the harvest from the last fishing trip.

MAISON LOUISIANNE   415-641-5920
P.O. Box 5704
South San Francisco, CA 94083

To experience Creole products that are tongue-tingling taste treats made from the finest Creole herbs and spice, you need to call **Maison Louisianne**.

Creole mustard is an old New Orleans flavor that, once tasted, will keep you from going back to yellow mustard. Mixed with yogurt or sour cream, it is a tasty dip. To add zest when basting meat, poultry, and fish, try the mustard straight. A natural substitute for regular mustard on sandwiches, Creole mustard mixed with apple cider and lemon juice makes an exquisite old New Orleans salad dressing.

If you are used to cooking with mayonnaise or salad dressing or putting them on sandwiches, your tastebuds need a wakeup call. Try the Creole mayo on a roast beef or ham sandwich and experience utter lunchtime satisfaction (extra spicy is available). In dips, tartar sauce, and cooking, the distinctive flavor will complement your favorite recipes. The "Shut My Mouth" Creole spice mix can also be ordered for those who enjoy an exciting meal. Sprinkle this blend on eggs or pasta and you won't be disappointed.

All Maison Louisianne products are available in various sizes suitable for individuals, families, caterers, or special events.

MANGANARO FOODS
488 Ninth Avenue
New York, NY 10018

800-472-5264
212-563-5331
Discount: Sample

If you have been searching for a unique assortment of authentic Italian foods, your quest is over! **Manganaro Foods**, established 1893, offers a vast selection of Italian edibles.

Your next genuine Italian meal should start with an antipasto plate. This includes a delicious collection of twelve of Italy's most famous appetizers, including sweet roasted peppers, caponata di melanzane (marinated eggplant and vegetable appetizer), seasoned sun-dried tomatoes, cacciatorini (sweet dried sausage), and more are offered.

There are extensive cheese and meat offerings. Pepato

is a typical "table cheese" that is also good in a sauce or for a special touch on homemade pizza. Fontina cannot be beat for a tasty melting cheese. Try Milano and Abruzzi sausages on the hors d'oeuvre plate, or the next time you stuff chicken breasts, use authentic Proscuitto ham with the Fontina cheese and you won't be disappointed.

Other products available include oils, vinegars, pastas, sweets, and coffees.

MIGUEL'S STOWE AWAY   802-244-7886
RR 3, Box 2086   802-244-7804 FAX
Waterbury, VT 05676

Authentic Mexican food knows no borders and is available straight from Vermont through **Miguel's Stowe Away**.

Originally started as an upscale restaurant, this company also caters to the specialty Mexican food market with a zesty group of products delivered to one's door. According to your tastes, choose from the three varieties of red salsa cruda—hot, regular, or mild. Accompany your choice with hand-cut White Corn Tortilla Chips; perhaps Blue Corn Tortilla Chips will add the color and taste you need. Both chips come in lightly salted or no-salt-added varieties.

Some of the newest products are red chile sauce and smoked jalapeño (chipotle) sauce. Combine these traditional tastes with the tortilla chips and the Sunday ball game or poker night will take on a new level of enjoyment.

MONTEREY PASTA CO.   408-646-5454
596 Lighthouse Avenue
Monterey, CA 93940

When you want to make a meal special by using fresh pasta, the **Monterey Pasta Co.** is well suited to supply your needs.

The company's pastas, raviolis, and sauces are made daily. Some of the fresh pastas are fettucine, angel hair, sweet lemon pepper, and rosemary garlic. Each will make one of your favorite regular dishes a gourmet treat. Delicious raviolis are prepared with a variety of stuffings—Gorgonzola, spinach & cheese, cheese, mixed vegetable, and smoked salmon.

The sauces include tomato and fresh basil, pesto, Alfredo, Parmesan primavera, creamy Gorgonzola, and lemon dill caper. Whether for your home cupboards or a gift, these sauces are delicious and healthy, and will not disappoint.

OLD SAN ANTONIO   800-972-3049
P.O. Box 10438   512-340-6615
San Antonio, TX 78210

If you enjoy a "hotter" touch to your seasonings or foods, **Old San Antonio** is prepared to put a little sweat on your brow.

Try one of their sauces. Chili con queso, which is a

blend of cheeses mixed with tomatoes, onions, and jalapeno peppers, is a perfect dip or nacho topping. Picante sauce, with jalapeños, tomatoes, and onions blended together with herbs and spices, is great on steak, eggs, and every type of Tex-Mex dish. Barbecue, enchilada, green chunky, and many other offerings are tailored to different foods for different times (from breakfast to a late-night snack).

Many Texan-fashioned jellies, mustards, and candies could be part of your future menu. Bluebonnet or beer jelly might liven up breakfast. Jalapeño or mesquite mustard is bound to improve many lunch sandwiches. And for that sweet tooth, experiment with a Texas Tranquilizer or two. These jalapeño jelly beans come cleverly labeled in a medicine bottle and are perfect for the person who likes "hot stuff." Just don't give one of their Jalapeño Lollipops to your two-year-old.

About Pasta

*P*asta is one of the oldest processed foods. Cheese runs a close second. Moreover, various forms of pasta originated independently in all corners of the world as ground grain and water were mixed into a "pastes."

Chinese records attest to the eating of noodles as early as 5000 B.C. Archaeologists have found evidence that during the Shang Dynasty, around 1700 B.C., a type of noodle was a staple in people's diet. The century before Christ, the Jews were stuffing pockets of pasta with lamb: hence the invention of ravioli. The Greeks had lasagnon *in Sicily before this.*

Different forms of pasta have been staples of many cultures. In some, their preparation was raised to the point of elegance. In the late 1700s, fashionable gentlemen gathered in macaroni clubs.

Macaroni became slang for anything "smart" and fashionable.

According to Karen Green, author of The Great International Noodle Experience, *the first American pasta factory was opened in Brooklyn, New York, in 1848 by a Frenchman, Antoine Zerega, using horses to mix and kneed the dough. In the twentieth century, millions of American immigrants made pasta a favorite.*

Americans only eat 10 pounds of pasta compared with the Italians, who consume 70 pounds per capita each year. But, the National Pasta Association says, Americans are enjoying more pasta each year (up from four and a half pounds in 1949). There are a least 100 shapes and an abundance of grains and flavors, ranging from buckwheat to lobster, that are used as ingredients. Unlike many other peoples, Americans often cover their pasta with heavy high-calorie fatty sauces. As a result, the misconception that pasta is fattening has become an ingrained culinary fallacy. Potatoes share a similar lack of understanding.

So live life to the fullest. East some pasta tonight. In fact, make some macaroni and cheese. You may be enjoying a casserole that dates back to the Roman and Greek eras.

PAPA LEONE FOOD ENTERPRISES, INC.
P.O. Box 6777   213-204-4300
Beverly Hills, CA 90212   213-204-4699 FAX

All-natural sauces. Just heat, pour, and enjoy. If this sounds interesting, contact **Papa Leone Food Enterprise** for a great taste sensation.

All seven Chef Alberto Leone Gourmet Entrée Sauces are meant to be heated—and then poured over cooked fish, poultry, shellfish, meat, vegetables, or pasta, to create dozens of elegant and delicious Continental dishes in seconds. So even if you only have a microwave, you can offer your guests something extraordinary.

The Italian country sauce is designed for steaks, chops,

and roasts. For a taste and wonderful treat, order the À L'Orange variety to complement a poultry dish. When there is a need to add life to a vegetable dish, the Zesty Vegetable is certain to bring your favorite alive. Whether you choose the Scampi or Fra Diavolo selections, your seafood will please the most discerning palate.

ROSSI PASTA   800-227-6774
P.O. Box 759   Discount: 10%
Marietta, OH 45750

If you are looking for something a little more exotic than spaghetti and meatballs, the handmade selections from **Rossi Pasta** might just be the thing.

This company uses only fresh ingredients harvested daily for its products (local garlic and free-range eggs, unbleached flour, and enriched spring wheat—the highest-protein baker's flour on the market). As far as selection, there are four pasta cuts: capelli d'angelo (thinnest, 1.25mm), tagliarini, linguini, and fettuccini (thickest. 6.00mm). Then they offer twenty-four varieties of these styles.

For variations on the standard pasta theme, you may want to try basil black pepper or parsley garlic fettuccini.

Perhaps seafood is on the menu and saffron linguini or lobster fettuccini is more appropriate.

Some of the other choices from Rossi Pasta include pimento fettuccini, citrus linguini, very chili linguini, and leek & onion linguini.

SAHADI IMPORTING   718-439-7779
187 Atlantic Avenue   718-643-4415 FAX
Brooklyn, NY 11201

For hard-to-find products of Italy, the Middle East, and the Far East, a good source is **Sahadi Importing**.

This company is best known for their Casbah product line, which offers traditional Middle Eastern cuisine items, such as wheat pilaf. Other healthy and tasty offerings include falafel topped with tahiti dressing, hummos, and tabouli.

For a true Italian flare in your cooking, try some of their Dal Raccolto products straight from Italy. Chianti and balsamic vinegar, sun-dried tomatoes, and porcini mushrooms would be the perfect ingredients to start a summer salad or unforgettable first course. In the main meal pesto alla Genovese, Italian red chili pepper sauce, and capers in balsamic vinegar are certain to liven up your favorite recipe or lead you to create a new masterpiece. Many of the Italian products also come in decorative packaging or decanters that make them perfect gifts.

For tastes of India, sliced mango pickle and other pickles make an ideal condiment for Indian, Middle Eastern, Greek, Armenian, and other ethnic dishes. The relishes come in five varieties—the genuine and regular sliced mango pickle in vinegar, mango pickle in oil, lime pickle in oil, and combination mixed pickle. All are perfect for the creative cook.

SAN ANGEL   802-253-8117
RR 2 Box 1390
Stowe, VT 05672

Do you want a direct line to Mexico's zesty products? Call **San Angel**.

One New England state known primarily for maple syrup and trees has a Mexican outpost. The tastes are truly Mexican but as San Angel's owners note, "We may be in Vermont, but we're not interested in hybrids" (Cal-Mex and Tex-Mex food). Some of the choices are salsa mexicana (which comes in mild, medium, and hot varieties), chipotle, which is hot and smoky, or the diablo, which will bring tears to your eyes, so it's not for those who don't like red hot.

Yellow and blue corn chips can be part of your order. For a perfect gift send the Sample Pack, including three jars of salsa, or the Fiesta Gift Box, which includes chips. Or if they really need a gift to tear up over, or if you have never experienced these delights, the chile ristras are a string of edible chile peppers ideal for home use or gifts.

SULTAN'S DELIGHT, INC.   800-852-5046 order
25 Croton Avenue   718-720-1557
Staten Island, NY 10301

For hard-to-find ethnic foods, particularly from the Middle East, Europe, and Scandinavia, **Sultan's Delight** is a definite choice for most anything.

Whether you need tahini (sesame butter) or the mix for traditional hummus or rice flour (for meghli), they are among the hundreds of items offered. Common and specialty spices, nuts, seeds, and flavorings that are hard, if not impossible, to find, are listed in one-quarter, one-, two-, and five-pound sizes. Particular products are offered in smaller, more convenient sizes.

Regional preserves, syrups, and juices are available to give any gathering a special exotic but delightful touch. You can challenge your guests to unveil the secret ingredient. Cheeses, pastries, and candies from all corners of the world will please a homesick traveler, exchange student, and probably even the neighbor who never seems to smile.

The Tastes of China

*T*he vast menu selection at a Chinese restaurant is delightful, but can be baffling! Egg rolls, a Pu-Pu platter, and fried rice are typically ordered. But these choices do not challenge the four distinct regional cuisines. In most cases the menu offers dishes from the four diverse Chinese styles, which range from spicy extra hot to plain, simple, but tasty! If enjoying the meal with a group, try a lot of different dishes to experience a traditional Chinese meal.

Szechuan and Hunan selections are among the most spicy-hot dishes. This cooking originates in southern China, where not only is the climate hot, but hot oil, hot peppers, and ginger are used liberally in most dishes. Hot and sour soup, hunan pork, and eggplant with garlic are zesty offerings that are sure to arouse a sleepy palate. Order them extra hot if your enjoy having a little steam come out of your ears while eating! The "hotness" is very different from the "burn a hole in your stomach" tendencies of some Mexican cuisine. When prepared correctly, the spiciness complements, but does not overwhelm the vegetables, fish, or meat in the dish.

The southeast region of China is the home of Cantonese cooking. In this most fertile region, fresh vegetables are available throughout the year and stir-frys are often the way they're cooked. Crisp, fresh vegetable dishes with red peppers, broccoli, pea pods, and water chestnuts combined with fried rice are the staples of this area. Cantonese entries are the most attractive selections for diet-restricted or health-conscious diners. In most cases, overwhelming spices and heavy gravies are not part of the Cantonese menu.

Shanghai cuisine is regarded as a haute cuisine. This eastern cooking style offers freshwater fish and seafood as fancy appetizers or the entrée. The fish is often roasted until the bones are brittle enough to chew. Various mustards and dipping sauces are generally offered as condiments. Unless your restaurant choice is in the Chinese section of your city, the traditional specialties, such as braised meats, eel in gravy, and pig's knuckles, will probably not be offered.

The northeast region of China is associated with classic or Mandarin cuisine. Rice is a luxury in this area as the growing season is short and dry. Carrots, turnips, cabbage, scallions, and garlic are most often used in combination with meat products. Deluxe vegetable dishes with chicken or beef in garlic sauce are typical of this region. This style of cooking is at the opposite end of the spectrum from Szechuan. All the entrées tend to be simple, but very tasty.

The traditional autonomy of each region has changed as people *in China travel and American tastes demand certain flavors. A restaurant may combine the spicy characteristics of Szechuan cooking in a Cantonese-styled stir-fry. Spicy shrimp and scallops combined with fresh vegetables and a "zip" of hot oil and ginger, for example. Yet the native distinctions can usually be found on most menus among the hundred or more offerings.*

So if you're planning to go out with three or more folks for some Chinese food, why not try all four regional cuisines? As the dishes are passed around the table compare then, note the distinct differences. Also taste the overlapping styles of the cuisine that make Chinese cooking a varied and tasty experience.

Note: *Do Not expect a fortune cookie at a traditional Chinese dinner in China or "Chinatown." The cookie and message inside were developed in San Francisco in the mid-nineteenth century as a way to pass secret messages in the Chinese underworld! The cookie is now strictly an American custom!*

SUSHI CHEF   212-772-6078
242 East 72d Street
New York, NY 10021

For those of you who enjoyed Sushi in the eighties, have your tastes "reborn" to its savory delights in the nineties! **Sushi Chef** will supply you with the ingredients for your home feast!

If you need shiitake mushrooms or konbu (giant sea-

weed grown north of Hokkaido), or perhaps rice vinegar or pungent Japanese dark soy sauce, this company is your source.

Gift packs are available! The Sushi Making Kit includes all the ingredients to create a sushi dinner or snack at home. The kit includes: pickled ginger, Japanese rice vinegar, Japanese sushi vinegar, nori (toasted seaweed), wasabi (powdered horseradish), and much more.

Over twenty items are offered which are well suited to the beginning chef or experimenting cook.

Chapter 5

Seafood and Fish

Since we have become more aware of the health problems due to high-saturated-fat diets, the eating of fish has increased markedly. Saturated fats (which are hard at room temperature) increase the bloodstream's *cholesterol* content, a substance found only in animal foods. Unsaturated fats (oils at room temperature) do not affect blood cholesterol adversely.

Recent discoveries show that groups of people who have a very high-fat "fishy" diet, the Greenland Eskimos, for instance, have a very low rate of heart disease. Fish oil is the reason. Heart experts are now recommending more frequent consumption of fatty fishes as one of many efforts to reduce the risk of heart disease. These include tuna and salmon steaks, sardines, mackerel, sable, whitefish, bluefish, swordfish, rainbow trout, and herring. The most preferable low-cholesterol shellfish are mussels, oysters, and scallops. Unfortunately, my favorite, naturally, is the highest in cholesterol—shrimp. Yet even one average-size serving has less cholesterol than one egg yolk.

Now, you can enjoy this chapter for two reasons. You

are ordering something delicious and something healthy. But don't be a complete saint and bypass the shrimp.

ALASKAN GOURMET SEAFOODS   800-288-3740
P.O. Box 190733   907-563-3752
Anchorage, AK 99519   907-563-2592 FAX

Experience the tastes of Alaskan seafood as if the fish were harvested in your backyard with **Alaskan Gourmet Seafoods.**

Try some of their delicious canned salmon—Hand-packed king, red, silver, and pink are selected to ensure quality and a delicacy that your family and friends will enjoy. If you are looking for something special for a birthday or holiday, they offer many gift boxes and baskets. All include canned fresh or smoked salmon, but you may also want to send Alaskan honey, wildberry jelly, and a salmon recipe book.

Entertaining, and you want to make an impression? Choose from the complete selection of Alaskan Seafood fresh from the icy waters to your doorstep. Vacuum-packed salmon, halibut, or combination packs with both. King crab legs already split for easy serving, snow crab, or four two-pound Dungeness crabs. Don't forget the smoked black cod or scallops.

ATLANTIC SEAFOOD DIRECT   800-227-1116
21 Merrill Drive   207-596-7152
P.O. Box 1128   207-594-2462 FAX
Rockland, ME 04841

Just set a date for dinner, call the 800 phone number, and **Atlantic Seafood Direct** will have your order at your home in time for the feast.

Try their Maine Shore Dinner, including live lobsters, steamer clams, clam chowder, and fresh chocolate truffles. The Land and Sea Supper might be more to your liking. This contains Atlantic lobster tails, fork-tender filet mignon, New England clam chowder, and fresh chocolate truffles. Both meals arrive with decorative placemats, napkins, and lobster bibs.

This company also offers smoked trout and salmon and many other irresistible lobster combinations. But don't miss the newest addition, "A Menu of Choice." In this directory of food items, you will find everything from single entrées to complete meals. Receive the New England Breakfast, including juniper bacon, Canadian bacon, ham steak, pancake mix, and pure Connecticut maple syrup. Or choose from the meat cuts, seafood combinations, or whole smoked turkeys or hams.

CAVIARTERIA, INC.　　　　　　　　800-4-CAVIAR
29 East 60th Street　　　　　　　　　212-759-7410
New York, NY 10022

When caviar is on the menu, skimping is never appropriate, and **Caviarteria** is the supplier you need to contact when presenting this delicacy.

All classes of Caspian, Beluga, and Sevruga caviar are offered. Their American sturgeon, whitefish, and salmon caviar take a standard hors d'oeuvres table to a pinnacle of delight.

For those folks on a budget, caviar is affordable! Once year, "Bottom of the Barrel" Kamchatka caviar is sold at reduced prices. The grains are partly broken, but the flavor "shouts" Beluga! The genuine Caspian sturgeon caviar is vacuum-packed and will keep fresh in the refrigerator up to one year.

Also, don't miss the smoked Scottish salmon, salmon steaks, Norwegian gravlax, fresh pâtés, foie gras from Israel and France, and more!

CLAMBAKE CELEBRATIONS   800-423-4038
5 Giddiah Hill Road   508-255-9610 FAX
Orleans, MA 02653

Dealing with a small company is okay...but one with a big heart is **Clambake Celebrations**.

You are invited to send one Lobsters-To-Go to anyone. But the Larger-Lobster-To-Go will certainly receive a larger response. A gift of four 2-pound-plus lobsters, packed in their steamer pot complete with accessories, is ideal for someone you closed that deal with, or when Uncle Bill drops in for a week.

Clambakes-To-Go offers a general seafood feast. Traditional New England clambakes include lobster, steamer clams, mussels, fresh corn on the cob, potatoes, and sausage and onions.

DUCKTRAP RIVER FISH FARM, INC.   800-828-3825
RFD 2, Box 378   207-763-3760 FAX
Lincolnville, ME 04849

If you are looking for top-quality naturally smoked Northeastern seafood and freshwater fish, **Ducktrap River Fish Farm, Inc.** awaits your order.

This company uses some of the most innovative smoking methods to bring out the optimum flavor, while keeping the product visually enticing. Sample their smoked eastern salmon, in which the traditional slow-cooking method creates a mild cure reminiscent of the finest European varieties. Smoked rainbow trout may be on the menu. These perfectly meated fish are netted, cleaned, and smoked *on order* and are waiting for your call.

Try one of the seafood/fish samplers that combine four distinct but complementary delicacies. Smoked mussels, presliced salmon, bay scallops, and trout fillets are part of this selection and can be the perfect alternative to a cracker and cheese plate or even serve as an entrée for a special meal.

 EKONE OYSTER CO.   206-875-5494
Star Route
Box 465
South Bend, WA 98586

Ahh—who doesn't love a *good* smoked oyster? **Ekone Oyster Company** is in business to bring you the finest smoked oysters available.

The fresh oysters are harvested directly from their own beds in the water of Willapa Bay. All the oysters are cultivated by a special long-line technique to keep the little creatures off the bottom so they feed better, feel better, and grow fatter than their relatives on the ground. This alone grows a superior oyster.

The two varieties offered are smoked in a fresh pack and the delights arrive in a convenient three-ounce can. The fresh pack contains freshly alderwood-smoked oysters that are rack-cooled and promptly sealed in a vacuum pouch to ensure that "from the sea" freshness to you.

HOMARUS INC.   800-666-8992
Dept. 101   914-666-8734 FAX
76 Kisco Avenue
Mt. Kisco, NY 10549

If you enjoy the delicate flavors of smoked fish and seafood, but your experiences have left your mouth "smoked," **Homarus Inc.** will eliminate this problem.

The seafood and freshwater fish offered surpass the requirements of the most demanding chefs, and will certainly be a treat for you or the recipient of the gift. Among the exquisite variations custom-cured Atlantic smoked salmon is "pastrami-styled" salmon sides. The Poivre-Lachs is smoked and coated with cracked peppercorns, garlic, and coriander. Some other versions include: glavlax prepared with dill, peppercorns, brown sugar, and salt; or with lemon dill, chrysanthemum petals, and tangerine rind (Hong Kong-style), or your choice. Yes, they will work with you to create classic smoked salmon or other smoked seafood.

Boned or unboned smoked trout is also available. Seafood specialties such as salmon roe caviar, eels, scallops, sturgeon, tuna, and more can adorn your dinner or buffet table.

HORTON'S NATURALLY SMOKED SEAFOODS
Gristmill Road   800-346-6066
P.O. Box 430   207-247-6900
Waterboro, ME 04087   207-247-6902 FAX

Remember the smoked mussels you bought at the grocery store, expecting a fun nibble but getting something that tasted like a soggy campfire! **Horton's Naturally Smoked Seafoods** from Maine will not leave your palate tasting of charcoal.

Among the smoked delights are Atlantic salmon, trout, mackerel, and catfish. Each is soaked in a specially balanced brine solution, which is crucial to the smoking process and the final succulent taste.

Also offered are salmon, trout and bluefish pâté. Whether spread on a bagel, as an ingredient in a cheese sandwich, or dolloped on cucumber slices, these pâtés take eating from being an ordinary experience to one that is at the pinnacle of pleasure.

When you order, don't forget the smoked shrimp, scallops, and mussels. You can mix them with a little oil, chopped garlic and pasta for a quick lunch or easy dinner. (Note sidebar on the history and methods of smoking meat and fish.)

Seafood 2001

*T*his isn't a piece on how a halibut grows to mammoth proportions and eats the Q.E. II. This is what Americans can expect from the seafood industry as we approach the year 2000.

According to the National Fisheries Institute (NFI), seafood consumption is on the rise. Presently at 15 pounds per capita, by the turn of the century, the industry will need to supply 20 pounds per person. The growth in the market is attributed to increasing concerns about health as well as the growing culinary appreciation by many people of the many types of available fish.

This new awareness has spurred the aquaculture industry—raising fish or shellfish in ponds or net pens. In 1988, nearly 8 percent of the seafood consumed in the United States was cultivated domestically. Today, America's fish farms provide 11 percent of the nation's total productions.

As we approach the twenty-first century, food manufacturing

technology will combine seafood proteins with other foods for more healthful products. There will be more products made from surimi, a popular pollack-based protein often used in grocery store seafood salad. Due to its unique protein binders, surimi can be used as a healthy ingredient in other foods and as the base for non seafood products.

Most Americans are only familiar with 10 or 12 species of fish and seafood. Yet across the country over 300 species are traded every day on the fish market. Orange roughy, tilapia, hoki, and kingclip will possibly be on your dinner table in the future, if they aren't already. Most consumers are not aware of how to handle, store, and prepare many of the seldom purchased varieties. With guidance from suppliers, the consumer will become more knowledgeable and discover the healthy, pound-for-pound cost-effective option that fish and seafood offer.

So be brave and follow the trend. The next time that "special someone" wants fish for dinner, find a good recipe for painted sweetlips fillets. The choice certainly has a nice "ring" to it.

JOSEPHSON'S SMOKEHOUSE & DOCK
P.O. Box 412   800-772-3474
Astoria, OR 97103   800-772-3474

From the tradition of gill-net fisherman comes the heritage of **Josephson's Smokehouse & Dock**; now their seafood is just a phone call away.

Only premium-grade chinook and coho salmon, sturgeon, and albacore tuna are selected for the fancy and smoked specialty seafood gift pack offerings. The many variations will make the receiver happy whether a birthday gift or thank you is needed. With the Hot Smoked selections the oysters, prawns, mussels, and shark will bring the tastes of the Pacific Northwest to your buffet table or special dinner.

For a surprise, instead of crackers and cheese, start with

smoked salmon and salmon pepper jerky. Or serve smoked shark jerky. The pickled seafood selections are equally enticing—pickled prawns, sweet & sour pickled salmon, and more. With all these treats, don't miss the salmon sides, sturgeon and salmon caviar, and other wonderful items.

LEGAL SEA FOODS   800-343-5804
33 Everett Street
Boston, MA 02134

"If it's not fresh! It's not Legal!"

Is your New England fantasy meal a Maine lobster with steamers? The Maine Event from **Legal Sea Foods** will satisfy your desires. This combination includes live lobsters, littleneck steamer clams, and creamy clam chowder. Even though the Berkowitz family, owners of Legal Sea Foods, may be busy assembling fish chowder or mussels au gratin for a presidential inauguration, they'd be glad to add fresh shrimp or calamari salad to your order.

The world-renowned Legal Sea Foods fish market ships many other delights: Swordfish, salmon, and halibut steaks can be on their way with a phone call! They are perfect for summer grilling or an elegant dinner party. Fresh sea scallops, cooked shrimp, or raw oysters can be an entrée or incorporated into the meal.

Other treats include the ready-to-eat smoked bluefish

pâté, creamy tartar sauce, and tangy cocktail sauce, or any of the succulent ready-to-heat items.

NELSON CRAB INC.   800-262-0069
P.O. Box 520   206-267-2921 FAX
Tokeland, WA 98590

For some of the tastiest seafood treats from the Pacific Northwest, let **Nelson Crab** supply your table.

Since 1940, Sea Treats have been given as gifts to lucky people throughout the country. These canned seafood delights are available in assorted gift packs for home cooking needs as well the needed thank you. Imagine the recipient's eyes when seeing Dungeness crab, smoked sturgeon, blueback salmon, Pacific shrimp, and more when opening the package. Many other tins are available for salads, hors d'oeuvres, and entrées. Columbia River blueback and chinook salmon are perfect to build a salad around, make a base for a spread, or alongside the cheese and cracker plate. Red sockeye salmon steaks will certainly be happily devoured by all invited.

RED-WING MEADOW FARM
P.O. Box 484
Sunderland, MA 01375

413-549-4118
Discount: Can of
Trout Nouveau

I bet you haven't had a "brook trout melt" recently! Or made a trout pâté for guests! **Red-Wing Meadow Farm** will change those culinary oversights.

Fresh brook trout (never frozen) are shipped the day they are harvested. Eight-ounce, ten-ounce, and twelve-ounce. portions are available. If you had a tough weekend angling, this could be the antidote for the lost bait. You can order the trout with the head or without, boned or unboned, and they will arrive for the feast you did not catch! Or for the non-angler, but creative cook, fresh trout is a tasty meal.

If you want prepared trout that can be used at your leisure, try the Trout Nouveau. The gourmet treat includes four- to five-inch canned brook trout. These arrive in any one of four flavors: mildly smoked, white wine, lemon/pepper and natural. Handsome two- and four-flavor gift sets are also offered. Whole lemon/pepper, and whole smoked brook trout are also featured and can arrive at your doorstep with only a phone call.

SANDER'S LOBSTER COMPANY   603-436-3716
54 Pray Street
Portsmouth, NH 03801

If you have been wondering what to get that special friend or relative who has everything, **Sander's Lobster Company** will solve your quandary.

All the lobsters are packed live, unless you request them to be cooked (parboiled). An assortment of sizes from "chicken," under a pound, to multiple-pound lobsters are available. Nothing could express your gratitude more than sending four pound-and-a-quarter delights to that special person in Arkansas.

To make a distinctive show of gratitude, send the Mini Clambake with 4 one-and-a-quarter-pound lobsters and 2 pounds of steamers. If you want to create your own package of seafood delights, simply call for the most recent prices and start "cookin" over the phone. But don't forget to order a couple of lobsters for your own dinner; after all, you deserve them.

SIMPLY SHRIMP   800-833-0888
7794 N.W. 44th Street   305-741-1944
Ft. Lauderdale, FL 33351   Discount: $1 per pound
                               off first order

If you are a seafood lover and want to experience Florida's finest, **Simply Shrimp** is your port of call.

For yourself, or a special gift, order the five-pound "Special Deluxe Package" of shrimp that will generously serve ten people. Each package includes a booklet of recipes along with storage and handling suggestions. The carefully balanced diet of the farm-raised succulent delights has turned them into the perfect treat for any occasion. They do not have the consistency of a bike tire, like some, and surpass the grocery store variety in both quality and taste.

Simply Shrimp also distributes stone crab claws. Native to southwest Florida and the Florida Keys, these adult hand-size claws are perfect for a casual dinner party or to thank your neighbor for watching your house while you were in Florida! The large claws arrive cooked and can be served with the complementary mustard sauce, or heated and eaten with warm drawn butter. Two or three claws are more than enough for a hungry seafood lover.

SPECIALTY SEAFOODS   800-645-3474
605 30th Street
Anacortes, WA 98221

For the finest northwestern Pacific smoked salmon, oysters, and related products, **Specialty Seafoods** is a perfect choice.

The offerings include handsomely packaged smoked fillets of sockeye, king, and North Pacific salmon to satisfy your

cravings, or are a perfect prelude to a special dinner. Smoked oysters (not like those the grocery store offers) could also complement the sitting.

A gift of seafood can also include healthy selections such as poached sockeye salmon fillets or two pounds of freshly frozen king salmon fillets. The fresh flash-frozen king salmon (known as chinook) is ready to cook and satisfy the appetites of four to six people. You can cook the fillets on the barbecue with dill butter or broil them with wedges of lemon. Either choice will be a crowd pleaser. Also try the ready-to-eat pâtés or canned salmon.

TOTEM SMOKEHOUSE   800-972-5666
Seattle's Pike Place Market   206-443-1710 WA
1906 Pike Place   206-861-8232 FAX
Seattle, WA 98101

For special gourmet seafood gifts from the Pacific Northwest, **Totem Smokehouse** has put together eighteen combinations that are certain to satisfy your palate.

Smoked sockeye salmon comes in a one-pound, four-ounce size for a dinner or buffet table, down to eight-ounce portions in individually sealed pouches. Each pouch serves two to four people. Premium Northwest oysters, or better yet, oysters and a sockeye salmon pouch are perfect for Dad and Mom on an anniversary, Uncle Louie because he is terrific, or even the boss!

Hand-packed canned seafoods are also available. Among some of the smoked delights are: salmon, sturgeon, oysters, clams, and salmon pâté. Other regional seafoods include sockeye salmon, Dungeness crab, tiny shrimp, and more.

Totem Smokehouse also offers no-salt, no-smoke sockeye fillets and center-cut slices already cooked and ready to eat.

WEATHERVANE   800-343-4000 Exc. ME
62 Badger's Island   207-439-0920 ME
Kittery, ME 03904

Throughout the world, the mere mention of Maine lobster brings a knowing smile to the faces of those who have ordered seafood from **Weathervane**.

Before ordering gifts, go ahead and send yourself a Downeast Clambake with one-and-a-quarter-pound lobsters, pounds of steamers, and fish chowder base. The package is complete with simple cooking instructions, two quarts of North Atlantic seawater (for cooking), lobster crackers and picks, placemats, and bibs. If you have the utensils, just order the Mainely Lobsters and however many lobsters you select will arrive ready to cook, at your door.

If North Atlantic seafood is your prey, choose from the Get Hooked selection. Prices are quoted daily on haddock, cod, salmon, smelts, swordfish, and more and can be delivered for your feast with only a phone call. Or if you want to send a peck of steamers (twelve pounds), or create your own oyster bar with scallops, jumbo shrimp, and Maine shrimp and more, you have the source.

Chapter 6
Fruits and Vegetables

Before you left your teens behind, there is no doubt that "Eat your vegetables" became a well-known phrase. After hearing this approximately a million times, second only to "Did you clean your room yet?" it tends to sink in. As adulthood approached, you probably began to like vegetables as your tastes changed, and fruit really was not all that bad, anyway. But just query a group of thirteen-year-olds about which vegetables they would abolish (if it were possible), and the list would probably sound very familiar from your younger days.

Pound-for-pound, dollar-for-dollar, fruits and vegetables are nutrition storehouses of vitamins, minerals, and dietary fiber. Good thing you developed a taste for them because, for the number of calories they supply, most contain a disproportionately large amount of essential vitamins and minerals.

When possible, always buy fresh fruits and vegetables. If you are concerned about the possibility of pesticide or fungicide residue, wash the produce in a bowl with two or three tablespoons of lemon juice or vinegar. This removes soil, bacteria, and water-soluble sprays.

If fresh vegetables and fruits aren't available, or the "fresh" produce is changing color or even supporting life of its own, frozen versions are nutritionally superior to canned. As a last resort use canned goods, realizing that many canned fruits have high concentrations of sweet sugary syrups, while canned vegetables may have added sugar or be heavily salted. By the way, I am no culinary saint, and always have canned goods stocked high on the shelves, but I use fresh produce whenever possible.

Unlike the other chapters, most of the companies in the Fruits and Vegetables section offer a limited product line. But you can be guaranteed that each business supplies premium fresh-quality produce. Due to their limited offerings, they have to deliver "the best," or they would be out of business.

BLAND FARMS   800-843-2542
P.O. Box 506
Glennville, GA 30427

Find out how easy it is to enjoy the irresistible flavor of fresh baby vidalia onions with a call to **Bland Farms**.

The jumbo, medium, and pee-wees are available May through June. These sweet-to-a-bite onions are in huge demand from sandwich makers and cooks. The baby vidalias are perfect for salads, casseroles, and certainly in your favorite dishes. They can be sent December to mid-April.

Along with the onions order from the dozens of country-style preserves and relishes. Or choose the Country Combo Basket, including marinated mushrooms, sweet pepper apple relish, hot chow chow, and mushroom tomato sauce, in a country gift basket, making a perfect birthday or anniversary gift. Or order vidalia sweet pickled onions, onion relish, and garden salad in a combo basket.

Bland Farms has more mouthwatering treats to offer.

BLUE HERON   800-237-3920
3221 Bay Shore Road   813-355-6946
Sarasota, FL 34234   Discount: Coupon

Friends, relatives, business associates, and even you can enjoy the sweet, juicy Florida fruits from **Blue Heron** with just a phone call.

Naturally, Florida oranges and grapefruit are offered. Navel and Valencia oranges, tree-ripened and sent fresh, along with pink, red, and golden white grapefruit can make this year's gift giving convenient. Perhaps you would prefer delectable Mineola tangelos (a hybrid of a tangerine and grapefruit) or delicious temples (a tangerine and orange hybrid). All will be welcome gifts from the "sunshine state" as the snow falls outside.

The orders are available in one-quarter bushel to full-bushel sizes. There are many fruit mixtures that can be delivered in authentic baskets to add charm to your gift. If you would like something more exotic, try the Tropical Fruit Combo: oranges, grapefruit, pineapple, coconut, and avocados. Or, try a selection throughout the year by joining the fruit-of-the-month club. They are offered in three-, six-, and nine-month schedules.

Most of the products from the Blue Heron are only available for limited seasons, but with their complete selection of fruits, you will be able to enjoy the juicy delights throughout the year.

Concerns About Produce

There are many conflicting concerns about how fruits and vegetables are grown, processed, stored, and delivered to your table. Worries about pesticides, fungicides, and fertilizers that are sprayed on during the growing period and waxes, colorings, and gases applied after harvesting are among the major concerns.

Organically grown produce from true *organic farmers is free of pesticide, fungicide, and fertilizer residue. Yet studies have shown that not all produce labeled organic has been grown without the use of such chemicals. Concerning the nutritional aspects, Jane Brody, in* Jane Brody's Good Food Book, *points out that there is no evidence that foods fertilized only with organic fertilizer are any more nutritious than foods grown in the usual, more commercial way. The nutrient content of a food has much more to do with its genes, the climate, when it is picked, and how it is shipped and stored than with the type of fertilizer used or soil on which it was grown.*

Waxes are applied to improve the appearance of many fruits and vegetables, prevent the food from drying out, and retard decay. The Food and Drug Administration considers them harmless but many nutrition experts disagree. For those who are concerned, wash-

ing and scrubbing the produce with soapy water may do the trick. Washing with a bowl of water and vinegar will remove water-soluble pesticides, fungicides, and dirt possibly containing fertilizer.

Color additives or fixatives are used to enhance the color or to stop discoloration. Some people are severely allergic to sodium bisulfite, which is used to stop mushrooms from darkening. Some fixatives used to enhance color (for instance, Florida oranges are greenish when ripe rather than bright orange) are possible carcinogens.

Gases used to speed the ripening process are similar to those produced naturally. If picked green, produce can be exposed to ethylene gas to commercially ripen the fruits and vegetables. This is very similar to the old trick of putting unripe fruit in a bag with an apple. The ethylene released by the apple ripens the other fruit. There is no known harm in this "gasing" of produce. The only drawback is that fruits ripened on the plant, vine, or tree have a higher amount of vitamin C than those picked green.

Careful cleaning is the best procedure to minimize possible carcinogens or allergic reactions. If you have children, the concerns must be amplified since many of these residues can affect proper growth and development. Please note the section titled "Children of a Lesser Food" in chapter 10 for more information concerning the effect on children.

THE CHILDREN'S CATALOG
CHILDREN'S HOME SOCIETY OF WASHINGTON

P.O. Box 15190   800-456-3338
Seattle, WA 98115   206-523-5727
                                  206-527-1667 FAX

If you would like to order premier produce from Washington State while benefiting a worthwhile cause, **The Children's Catalog** is a perfect choice.

A gift from this catalog is a gift that gives twice: once to the recipient of the produce and again, most importantly, to

the children of the Children's Home Society of Washington, founded in 1895. The CHSW offers counseling for parents, abused children, and an extensive range of support programs involving adoptions and foster care in the state of Washington.

All the gifts are seasonal. Red delicious and Granny Smith apples, along with bosc pears, are available October through May. Genuine Washington state Walla Walla sweet onions are offered for a short time in July. These soft-ball-size onions are so sweet that you can eat them like an apple, with no aftertaste.

Plans are being made to expand the product line. Even though The Children's Catalog is now limited in the number of produce items offered, the efforts made with the profits from the sales are very worthwhile. This alone should place them high on your list of mail-order food companies to patronize when their produce is available.

DELFTREE CORPORATION
Hidden Valley Road
P.O. Box 460
Pownal, VT 05261

800-243-3742 order
413-664-4907 Corp

Since ancient times, the Japanese have prized the shiitake mushroom for its exquisite flavor and tender, beef-like texture. Now **Delftree Corporation** offers the delicious, robust mushrooms for your cooking experience.

These organically grown mushrooms have been served in many fine restaurants for years. Now you can include them in your own prized recipes. Shiitake mushrooms can be marinated and broiled as a perfect topping for steaks or roast beef. A stir-fried meal will take on an exciting new personality with these plump, exquisitely flavored additions. Stew them in a sauce, or serve them fresh, with dips, or in a salad.

DIAMOND ORGANICS	800-922-2396
P.O. Box 2159	408-662-9714
Freedom, CA 95019	

Today only 1 percent if the fresh produce grown in this country is organically grown, and **Diamond Organics** can help you receive some.

Diamond is a purveyor of organically grown specialty lettuces, greens, herbs, roots, and fruits for top-quality restaurants and hotels. Now that quality is available through to-your-door service. Enjoy or send lettuce, mesclin salad greens, or the assortment of specialty greens including:

arugula, red mustard, Italian blue kale, and a dozen more. Fresh herbs and edible flowers to enhance your favorite recipe or to experiment with are ready for shipment. Many tips and information numbers on these less used delights are offered.

Cultivated and wild hand-picked mushrooms for a special gift include: shiitake, pom pom blanc, baby blue oyster, and more. Pick from the winter squash and you will find acorn, chestnut, Hokaido, and red kuri to enjoy in soups and casseroles; baked, boiled, steamed; mashed, cubed, or eaten right from their skins. Do not overlook the fantastic fruits. Apples, pears, citrus, and many more fruits are available, and they are organic.

FOLSOM FARM   800-252-6129
Rt. 3, Box 249
Glennville, GA 30427

If you use onions only for cooking, you have not enjoyed the delicious treat of "sweet large onions" from **Folsom Farm**.

This product is unlike the ordinary onion that brings unpleasant odors to your breath. They are not the hot onions that you may be accustomed to, but a delicious, mild, sweet onion known as the vidalia sweet onion. It is grown in a small area in southeastern Georgia. Many farmers in different locations have tried to duplicate the mildness but have not come close due to differences in climate and soil texture.

The onions come in ten-, twenty-five-, and fifty-pound bags, in jumbo, medium, and peewee sizes. Other onion products to please the onion lover in your house include onion relish, jelly, and vinaigrette. Combination gift packs are available.

Please note that the onions are only shipped between May 1 and June 30.

FRANK LEWIS'S   800-477-4773
Alamo Fruit
100 North Tower Road
Alamo, TX 78516

For luscious gift fruits or your own fruit bowl, **Frank Lewis's** has a top-quality selection.

This company excels in supplying unique fruit packs containing one or more of their offerings. Most of the products are seasonal, so it is wise to check before ordering. Among the harvests are Royal Red Delicious apples, Royal Fuji apples (a rare gourmet delight), and plump, juicy Royal Comice and Red Anjou pears.

Other offerings include Royal Asian Pears, which are large round fruit with a firm, munchy texture and the refreshing pear flavor we all enjoy. Royal ruby red grapefruit, canned ruby red sections, and juice can also make your gift-giving quick, easy, and very tasty.

All Frank Lewis's orders are shipped fresh from the orchards to ensure that they arrive crisp and juicy. Gift packs containing sun-dried fruits and dates are also available.

G.I.M.M. DRY YARD   916-795-2919
5030 Wolfskill Road
P.O. Box 1016
Winters, CA 95694

For some of the best-quality California dried fruits, be sure to contact **G.I.M.M. Dry Yard** with your next order.

This company hand-cuts and sun-dries selected California fruits and sends delicious mixed-fruit boxes or single-fruit boxes that are perfect gifts and treats. The dried fruits offered are apricots, prunes, pears, peaches, black figs, and nectarines.

The dried fruits are shipped in a variety of forms. Cut or whole, mixed or single, pits or pitted, in almost any combination, including twenty-five pound bulk cases. Many of the orders are accented with fresh California nuts and make a perfect, healthy gift for family or friends.

G & R FARMS   800-522-0567
Rt. 3, Box 35A   912-654-3030 FAX
Glennville, GA 30427

For year-round gift-giving needs filled straight from a working farm, call **G & R Farms** with your order.

This company has the famous and delicious vidalia onions throughout the year due to the recent construction of a controlled-atmosphere storage facility. Buy for yourself or send medium and jumbo packs in ten- and twenty-five-pound shipments. For that person with a green thumb, bundles of plants are available, and why not send something to be enjoyed now, like one of their farm samplers. Sweet onion vinaigrette will make a superior salad out of a good one.

For a larger selection, send the Ring the Dinner Bell. This large, tasty sampler includes sweet onion relish, onion

barbecue sauce, onion jelly, Georgia peach preserves, country-style pecan syrup, one pound of onion ring batter, and two pounds of stone-ground speckled heart grits. This wonderful collection of goodies is bound to be well received and completely enjoyed.

H. F. ALLEN COMPANY   800-444-9540
P.O. Box Drawer F
Vidalia, GA 30474

For some of the finest seasonal produce from the heart of Georgia, contact **H. F. Allen Company** with your order.

All of the products are seasonal, but while they're available, you'll get the best and the freshest. Shipping of vidalia onions starts in May and continues until the supply is exhausted. These sweet delights come in ten-, twenty-five- and fifty-pound orders. Small, medium, and jumbo, Vidalia onion plants are available early in November through March.

Vidalia sweet potatoes are another seasonal treat. They come in ten-, twenty-five- and fifty-pound orders and are perfect for the barbecue or as a gift to that person who can never have enough. Mix or Match Gift Packs provide a choice of onion pickles, garden salad, mustard relish, etc. The gift packs make an ideal addition to the wonderful seasonal produce.

HADLEY FRUIT ORCHARDS   800-854-5655 Exc. CA
P.O. Box 495   800-472-5672 CA
Cabazon, CA 92230   Discount: Sample

If you are a fan of dried fruits and premium-quality nuts, **Hadley Fruit Orchards** welcomes your orders.

They offer seven delicious types of dates conveniently listed in a "date comparison chart" so you can match your needs with the right date! The uses and tastes of each type of date are described. If you want very sweet, medium, snacking, or dates for pitting and chopping, this chart is very helpful. Perhaps you stuff dates or need a creamy or chewy fruit for cooking; the information provided will ensure the proper choice.

Many delicious mixtures of "munchies" are also available. Tropical Isle Mix combines pineapple, papaya, and banana chips with Spanish peanuts, Brazil nuts, ruby and golden raisins, and date nuggets for a taste of paradise. The Original Trail Mix blends delectable ruby raisins, sunflower and pumpkin seeds, peanuts, cashews, and almonds for a great pick-me-up treat.

HAZEL DELL MUSHROOMS   408-724-0673
P.O. Box 281
Watsonville, CA 95077

Whether you are a gourmet cook or just want to experiment with nature's exotic edible mushrooms, call **Hazel Dell Mushrooms** soon.

On a forestlike controlled climate farm, the conditions exist for producing the tasty tree oyster and shiitake fungi. These marvelous tree mushrooms are perfect in dishes where their taste is the center of attention. Using both varieties in one meal is certain to leave your guests wondering about your source.

The two varieties offered are Tree Oysters and Shiitake; each has its own unique woodsy flavor and distinctive texture. The Tree Oyster is crisp and tangy when raw but mild and subtle when cooked. Shiitake are wonderfully buttery in a salad, casserole, or stir-fry.

MISS SCARLETT   800-345-6734 Exc. CA
P.O. Box 1488   415-340-9600 CA
Burlingame,CA 94011

For some of the most outstanding condiment jellies, vegetables soaked in exquisite marinades, and fruits bathed in liqueur, contact **Miss Scarlett**.

The Red Pepper Jelly is their signature product and definitely worth a try. This medium-hot jelly is delicious when served with cream cheese, on crackers, will make a perfect addition to lamb, and is unbeatable when broiled in a fresh

---

pear or peach half and served with a meat entrée. Don't miss ordering the Ginger Pear Jelly, either, for exciting cooking experiences or as a gift.

If quality marinated cooking and salad olives, mushrooms, artichokes, and much more are lacking in your cupboard, you can now fill that void. Start with spiced garlic or marinated Italian olives. Include an order of marinated miniature Eggplants and miniature corn. To top the gift or self-indulgence off, incorporate a jar or two of peaches in brandy or the very popular California marinated vegetable salad.

MISSION ORCHARDS   800-526-0900
3501 Taylor Drive
P.O. Box 8505
Ukiah, CA 94582

When a show of appreciation is needed, for colleagues, clients, or customers who value superior quality, **Mission Orchards** is ready to fill your order.

Whether you are interested in fresh Fruit-for-All-Seasons clubs or gourmet specialties, this company has the

right combination for your gift-giving needs. Enjoy delectable juicy crown comice pears, red mission apples, and succulent California navel oranges. Some packs include the tasty addition of Gouda cheese, dried fruit, or mixed nuts. Many other fruits round out their offerings.

For entertaining and gift giving, perhaps smokehouse salmon, country smoked turkey, or spiral cut ham are needed. The specialty store entices you with pickled green beans, hearts of palm, and a dozen cans of the perfect tongue-in-cheek gift, Spam. Don't miss the dried fruits, jelly beans, nuts, cookies, and delicious deserts. A six- or twelve-month Dessert Club is a perfect memento, or wonderful self-indulgence throughout the year.

MISTER SPEAR   209-464-5365
P.O. Box 1528   209-464-3846 FAX
Stockton, CA 95201

If you're looking for the best, you've found it. Today, **Mister Spear** has simply the most succulent asparagus available anywhere.

This claim would be hard to make with nothing to support it, but the company comes through with nine gift packs featuring three varieties of superb asparagus. Colossal Green spears are canned and are the only product that is available all year. Fresh Jumbo Green and Fresh Large White spears are seasonal, and are

---

usually available only from March to early June. Weather plays a large factor at the beginning and end of the season.

In addition to the three asparagus products, this company may be expanding their product line. Brussels sprouts may be the newest offering, so be sure to keep updated in the coming season.

ORANGE BLOSSOM MARKET   800-624-8835
5151 S. Orange Blossom Trail   407-851-2601 FAX
Orlando, FL 32839

For the luscious tastes of Florida citrus, **Orange Blossom Market** can fill your order.

What could be a nicer gift or way to fill the fruit bowl than with a box of pink seedless grapefruit, Valencia oranges, or a combination of both. These delicious fruits come in one-quarter to full-bushel sizes. Perhaps the Sunshine Sampler would be a big hit. This gift contains eight large oranges, each in its own wrapper, plus delectable Coconut Patties and a three-ounce jar of orange marmalade. The box can be ordered with nine oranges only.

Perhaps a juicy reminder is needed throughout the

year. The Season Fruit Club is available in three-, six- or eight-month plans. Just order once and they will do the rest. Some of the succulent delights you or the lucky recipient will receive are Minneola tangelos, Temple oranges, Murcott homey tangerines, Valencia oranges, and more. You can send from three half bushels to eight full bushels per order each year. European shipments are even available, and there are many other selections.

PEZZINI FARMS   800-347-6118
P.O. Box 1276
Castroville, CA 95012

For the connoisseur of artichokes, or those who have yet to delight in this wonderful vegetable, **Pezzini Farms** is waiting for your call.

Castroville is considered "The Artichoke Capital of the World," so to find the freshness offered by this company is probably not surprising. They will ship their fresh, flavorful delights to your door for a feast with friends, family, or a special dinner that needs something fancier than peas. They also present steamed and French-fried artichokes.

To make your meal even more enjoyable, serve one of their dips created just for artichokes. Drawn butter is still a great dip to use, but nothing complements a freshly cooked artichoke better than dips conjured to match their flavor. Among your choices are pesto, lemon-dill, onion herb, or garlic Dijon dip, any of which will make the butter seem an also-ran. Some of the many other products include dried fruit and gift packs.

PIKLED GARLIK   408-372-7944
P.O. Box 846
Pacific Grove, CA 93950

"If It's Not Pikled—It's Just Garlic." Garlic lovers, take note of **Pikled Garlik** for home supplies and gifts.

This company has taken a basic ingredient in cooking, garlic, and pushed it to the pinnacle of the culinary ladder. Fresh peeled garlic cloves are combined with natural spices and herbs in an appetizing pickling liquid to create these taste and cooking sensations. The cloves come in mild, jalapeño, and red chili, and will surprise your palate and become a regular addition to your favorite recipes.

The healthfulness of garlic in your diet is only another reason to incorporate this company's pickled delights into your cooking. Try them sliced in a sandwich or salad. Dice them into sauces, casseroles, and dips. When you're making everything from homemade dressing to broiled fish, these tasty surprises will delight.

PINNACLE ORCHARDS   800-242-2485 Corp.
Corporate Sales Division   800-759-1232 Order
441 S. Fir, Room 66
Medford, OR 97501

When you want to send produce, whether for an individual or twenty-six hundred employees, choose the company that Nike picked—**Pinnacle Orchards**.

If a gift of fruit is on your list, select from the sumptuous red or green comice pears, red or golden delicious apples, golden sun-ripened oranges, luscious ripe peaches, red bartlett pears, and so on. All come conveniently packaged in various sizes, some in baskets and other combination fruit gifts. If you want to give more than fruit, other additions include bricks of cheddar cheese, roasted or smoked almonds, dates, chocolate truffles, and a variety of other snacks to please the recipient.

After sending the gift, treat yourself to an order of Poached Pears Eleanor. Poach and peel some bosc, or comice pears for a few minutes in apple juice or white wine (a sweet wine is needed here, perhaps a French Sauternes or German Berre Alsace). When tender, place in dessert dishes and top with paradigm chocolate fudge or vanilla caramel sauce. Your palate will ascend to heaven on the first taste.

Garlic Folklore

*I*n antiquity a common name for garlic was "cure-all," and through the centuries this simple bulb has maintained its reputation for healing everything from simple infections to high blood pressure and tuberculosis.

The ancient Greeks and Romans prescribed garlic for hundreds *of specific ailments. In the sixteenth century, Parisians were promised good health year-round if they would eat garlic with fresh butter during the month of May. The British actually used garlic to control infection during World War I, and the Russians used it to control flu epidemics.*

There is probably more lore about the tiny clove of garlic than about any other food. Folk medicine says that a cold will surely be cured if one rubs the soles of the feet with cut cloves of garlic. For a toothache, there are two schools of thought: one, that a sliver of garlic placed in the cavity of the tooth will relieve the ache: the other says that the sufferer should place a slice of garlic in his or her ear.

History indicates that King Henry V was anointed at birth with wine and garlic because it was believed that garlic on a baby's lips served as a stimulant and antiseptic. But Millin, writing in 1792, praised garlic as a preventive against the plague, and Bernardin de Saint-Pierre recorded that garlic cured nervous maladies.

Scientific proof or no, the belief that garlic can "cure what ails you" persists and in fact, seems to be on the increase. But whether medical science ever reaches any definite conclusions about the curative powers of garlic seems of little consequence to true garlic lovers. Most important is the magic which garlic performs in the kitchen. The delicious flavors that result from blending garlic with other foods can only be described as pure witchcraft.

SANTA BARBARA OLIVE CO.   800-624-4896
P.O. Box 1570   805-688-9917
Santa Ynez, CA 93460   805-686-1659 FAX

If olives are round things that are either green or black, you have not experienced olives from **Santa Barbara Olive Co.**

Try an order of green spiced olives. Some of the dozen or more offerings, all using

Berrouni olives include: Country style—in a country-herb-and-spice brine; Italian Style—with Italian spices, bay leaves, mild red peppers, and garlic; Mexican Style—with red chiles, garlic, and a secret family spice blend. Even plain green are available, but there is nothing plain about the superb quality.

If you have never had a hand-stuffed olive, now is your chance, with garlic, mushroom, anchovy, and almond stuffings as part of the selection. But if the only olive you enjoy is the pimento-stuffed martini olive, you need to order a Gourmet Oliver Sampler with four five-ounce spiced olive jars and an extra-virgin olive oil for a starter. Send one as a special gift and you are sure to get rave reviews.

SPHINX DATE RANCH   800-482-3283
3039 N. Scottsdale Road   602-941-2261
Scottsdale, AZ 85251   602-941-1840 FAX

When the gift calls for fresh dates or dried fruits, **Sphinx Date Ranch** has a solution for you.

For that special client or person you are not sure how to say thanks to, one of the Royal Gold Gift Boxes is perfect. Choose a box of the Sphinx's signature dates, the Select Sphinx Medjools or Medjools and Pecan-Stuffed Medjools, to show your gratitude. Maybe the occasion requires the Gourmet Sampler. One of the selections offers Chocolate-Dipped, Walnut-Stuffed, Coconut-Rolled, and standard Medjool dates.

Also offered is a wide variety of one-pound bags of nuts. Whether for a gift or personal use, a bag of almonds, macadamias, or English walnuts is sure to be a winner. Many dried fruits are also available. Apples, nectarines, and peaches are just a hint of the selection, which also includes Australian glacéed and tropical dried fruits.

SUNSHINE GIFT FRUIT SHIPPERS, INC.

P.O. Box 8   800-932-6543
Goulds, FL 33170   305-245-9222 FL
                                                    305-245-1502 FAX

For the superb taste and flavor of Florida's Indian River oranges and grapefruit, **Sunshine Gift Fruit Shippers** is ready to take your pick.

Among the seasonal delights offered are ruby red grapefruit, navel oranges, honeybell tangelos, temple & Valencia oranges, and honey tangerines. Mixed oranges & grapefruit are packed in tasty one-fifth to one-and-a-half bushel gift combinations. The All-Grapefruit Pack of plump ruby red or sweet white seedless are perfect gifts for that absent relative or business colleague.

The exclusive gift might include the Champagne Basket sent in a reusable wicker hamper with your choice of luscious Florida citrus. Or the Palmbo Basket in a colorful handwoven, sturdy straw basket filled to the brim with sweet and juicy Florida oranges and grapefruit. Both are a clever way to brighten a special day.

Note that the deliveries of Sunshine Gift Fruit Shippers fruits and baskets are seasonal and European parcels are available, so check the Fruit Calendar in their brochure.

TIMBER CREST FARMS   707-433-8251
4791 Dry Creek Road   707-433-8255 FAX
Healdsburg, CA 95448

Timber Crest Farms is a brand name that people look for when seeking the finest dried fruit or tomatoes, and they are waiting to provide you the very best.

Throughout the catalog you will find delicious and wholesome dried fruits, nuts, and gourmet foods that are a

great addition to any meal. Over twelve varieties of dried fruits are ready for cooking or the snack table at your next gathering. Pears, peaches, and apricots are part of the list, with more exotic choices like starfruit, papayas, and mangos for that special touch.

Sonoma dried tomatoes are also a featured product. Other offerings include dried tomato pasta sauce, dried tomato chutney, and dried tomato salsa. Nuts, gift baskets, and a collection of dried fruit and tomato products are ready to be shipped to garnish your table.

Chapter 7

Meat, Poultry and Game

What would a barbecue be without sizzling steaks, chicken, hamburgers, and hotdogs? I am not quite sure, but it probably wouldn't prompt your neighbor to return your chain saw.

This chapter contains companies offering top-quality barbecue food items and special cuts and types of meat, poultry, and game that you may not be able to buy locally. Many of the tastes are delightfully unique and delicious. Try some moose meat, for instance. Then make an extra sandwich, present the treat to a coworker, and watch the reaction. Surprise! Now you know what to get your colleague for Christmas.

BUCKMASTER   800-252-4692
Capoli Ranch   319-538-4888
Lansing, IA 52151

Gourmets around the world treasure venison delicacies; now you can have them through **Buckmaster**.

From cooking to slicing, to a sandwich on the trail, venison will make the meal. For the hunter or trail "hound" of the family, send the Capoli Venison Gift Box. This venison assortment contains a ten-ounce original and Old World summer sausages, deerslims, and a packet of peppersticks. The deerslims are spicy meat sticks that are perfect for snacking, and the peppersticks will be a delightful surprise, no hint. Ordering single items is not a problem.

If the hunted have alluded the hunter, a gift of country breakfast sausage or country bratwurst will soothe the loss and fill the hunger so one can go out again. Or these treats can be grilled, simmered, and placed on a buffet table, with your own sauce, or plain to offer a twist to cheese and crackers.

BURGERS' OZARK COUNTRY CURED HAMS
Route 3, Box 126
Highway 87 South   800-624-5426
California, MO 65018   Discount:
                                 Telephone Special

For the widest variety of naturally cured whole hams, sliced, smoked or glazed, **Burgers'** can supply superior quality for a reasonable price.

All the selections are aged for six months to one year

(note the sidebar). Some hams are cured with salt, spices, and herbs, specially attic-aged, sugar-cured, or smoked. This is just part of the offerings. Preparation of whole and half hams can be spiral spiced, uncooked, and vacuum packed, or cooked. Half hams are also offered cooked plain or with honey glaze. The cured center-cut steaks, aged either four or six months, or one year (your choice), are perfect for an evening meal or afternoon barbecue.

Among the assortment of other mouthwatering offerings: Ozark beef and pork sausages, smoked chicken or pork (ideal for a stir-fry), country ham bacon, and country rope sausages. Maybe you should add the beef or the oven-ready honey, sorghum, and brown sugar–basted turkey to your menu.

Many other offerings, from pheasant to gift packs, are just a phone call away. "Food-of-the-Month" deliveries can also be scheduled by Burgers'.

CABELA'S   800-237-4444
812 13th Avenue
Sidney, NE 69160

Each of **Cabela's** choices of smoked fish and game has been selected to satisfy the most discriminating palate. Why not yours?

If you want to treat a hunter, angler, or outfitter with some of the best hickory-smoked meats, fish, and cheese, try the Sportsman's Smoked Snack Sampler. You get beef, hot beef, venison, and buffalo sausage in an irresistible pack that also includes cheddar, hot pepper, and onion cheese. The icing to the gift is with the spicy meat sticks: beef, buffalo, and venison.

Choice meats ready for the family dinner table or special holiday are a perfect addition. Smoked turkey, wild turkey, and ringneck pheasant are just a hint of the entrée selections. Hickory-smoked duck and smoked ham would make any gathering more festive. Don't miss the turkey and beef jerky strips or the Wold Country Assortment containing duck, pheasant, wild boar, antelope, and Cajun brats (hot-doglike sausage).

CHOPS OF IOWA, INC.   800-252-4692
105 SW Locust   800-242-4692*
P.O. Box 190   515-758-2317 FAX
Earlham, IA 50072

For a great variety of meats, including specialties not found elsewhere, **Chops of Iowa, Inc.** is waiting to take your order.

Choose from the assortment of pork chops that are unlike the average grocery store meat. The Chops of Iowa chop is thick and perfect for hearty appetites and the most discriminating palate. Try the smoked Windsor chops for a

---

* Use this number outside the forty-eight contiguous states.

taste treat with your guests and gift recipients. These subtle hamlike chops are a choice entrée when grilling for the family. If you are looking for the classic roast pig, your extraordinary celebration dinner is prepared and completely cooked. Just heat and serve with an apple in its mouth.

If pork is not your choice, there are numerous delectable beef cuts. T-bone steaks, Iowa strip sirloins, and Iowa's best filet mignons, for example. Don't miss the Tailored Choices for memorable gifts or anniversary celebrations. Cakes, cheese, and much more are also available.

CIMPL MEATS   605-665-1665
1000 Cattle Drive
Box 80
Yankton, SD 57078

In the tradition of Old World Sausage Makers, **Cimpl Meats** brings the taste home.

Choose one of their savory gift pack assortments, which make an excellent holiday, birthday, or thanks-for-the-help gift. The Old World Pak is a taste experience four times over: New England Brand Sausage, cervelat, Proschke Brand Sausage, and German bologna. Or make your entertaining easy with the Holiday Special and enjoy everything from sausage on rye to cheese and crackers. This pack is ideal for the starving college student who needs a quick nutrition fix between tests.

To get someone's attention, a wink might do but ten choice rib eyes will catch their eye. For broiling or the barbecue, six T-bones are sure to please. Fillets, combination boxes, and special cuts are available upon request.

COOKING MASTERS   708-697-1300
220 N. Spring Street   800-252-5438
Elgin, IL 60120   708-697-1353 FAX

If you are looking to purchase food entrées for a large dinner party, business meeting, or even if you have a sizable family and own a freezer, **Cooking Masters** is your source for ready-to-heat gourmet goods.

The Traditional Stuffed Breast (chicken) Entrée is a wing joint, boned, the skin removed, then breaded and pre-browned. The four variations are Kiev: filled with 93 score AA grade butter, chives, herbs, and seasonings; Cordon Bleu: filled with Canadian bacon and Swiss cheese; Apple and Almond: filled with apples, croutons, toasted almonds, cinnamon, and special seasoning; Marco Polo: filled with broccoli and diced Canadian bacon in a sharp cheddar cheese sauce. All are lightly breaded and heat in about thirty minutes.

Some of the other offerings include twice baked potatoes, hot hors d'oeuvres, veal cordon bleu, Cornish supreme, puff pastry entrées, and more. Note that these prepared foods are shipped in large quantities, ready for a large gathering or to freeze and use over a period of time. The Traditional Stuffed Breast, for instance, serves sixty. Many of the other dishes are shipped in more reasonable quantities but would still serve a large family for a few meals, or could even be split with a neighbor.

DANIEL WEAVER COMPANY   800-WEAVERS
15th Avenue and Weavertown Road   717-274-6100
P.O. Box 525
Lebanon, PA 17042

Had your fill of bologna? Well, the **Daniel Weaver Company** will change your mind with just one taste.

Since 1885, this company has churned out the savory, ready-to-eat, all-beef smoked sausage that started as a local favorite but is now enjoyed on every continent. The "cold smoke" cured, 90 percent lean bologna comes in the traditional Lebanon, sweet, and beef roll-style. These varieties are certain to convince the bologna sandwich maker that they have lived for years without tasting the real thing.

Direct from Amish Country come other meaty delights. Canadian bacon and wood-smoked ham and bacon are perfect for a breakfast, lunch, or dinner. Wood-smoked turkey is ideal for your holiday buffet. The dried, or "chipped," beef will bring your favorite recipe to life. If you want to enjoy more than one taste treat, choose from the many combination gift packs.

FRED USINGER, INC.   800-558-9998
1030 North Old World 3d Street   [414] 276-9105
Milwaukee, WI 53202

At **Fred Usinger, Inc.**, "America's Finest Sausage" is more than a slogan. It is an ideal and family pledge by the *wurstmachers* who have been following the original family recipes since 1880.

This company offers twenty-one gift packs to please the fussiest *feinschmeckers* (gourmets). For your next gathering, Gift Number 2, the Crowd Pleaser, is bound to be a hit. It includes braunschweiger liver sausage, all-beef summer sausage, weiners, Strassburger liver sausage, mortadella, beerwurst, yachtwurst, aged Wisconsin cheddar, and snack rye bread. Add your choice of mustards and your hors d'oeuvres are complete.

The other gift packs contain numerous smoked delights and cooked meats. Smoked pork shoulder, slab bacon, Canadian bacon, and the ambrosia fudge assortment are only a few of the items that can adorn your table. Smoked holiday hams and luncheon loafs for tasty meals or superb sandwich makings are also included in their product line.

GRIDLEY'S FINE BAR-B-Q   901-794-5997
P.O. Box 22405
Memphis, TN 38122

If you haven't had barbecued meats that have been slow-cooked for fifteen or sixteen hours to bring out the best quality, you have yet to eat the finest and should contact **Gridley's Fine Bar-B-Q** immediately.

The same top-quality barbecued ribs that can be ordered at Gridley's Restaurant can now be on your dinner table for an authentic barbecue meal. The biggest challenge for the company was to produce the *best*, if not perfect, sauce. They have succeeded and offer mild, spicy, and hot versions.

The two main offerings are a Rib Box, which includes three slabs of top-quality ribs, bread, and the sauce of your choice. This feeds six hungry people. The Bar-B-Q contains five pounds of barbecued meat, and bread and sauce, and feeds fifteen to twenty folks. The three varieties of sauce come in cases (fifteen containers per case), in eight- or thirty-seven-ounce containers so that you may experiment with the "perfect" sauces at home.

HONEY BAKED HAM CO.   800-343-HAMS
105 Green Street
Marblehead, MA 01945

If you haven't feasted on a **Honey Baked** spiral-sliced ham recently, your time has come.

People coast to coast love Honey Baked Ham. The hams are covered with a tantalizing secret spice glaze that seals in the natural goodness. Smoked and slowly cooked for over thirty hours, these entrées are ready to eat, can be warmed for a buffet, or given as a gift.

Honey Baked Ham is sold in more than two hundred Honey Baked stores across America. Call for the nearest location to receive a "spiral cut" ham soon.

KRABLOONIK COUNTRY MEAT AND SPECIALTY
19101 Highway 82   303-963-9013
Box 28607
El Jebel, CO 81628

Setting the standard for savory smoked meat and game sausages, **Krabloonik Country Meat and Specialty** has established its reputation by supplying high-quality country delicacies delivered to your door.

Try their selection of Natural Mountain Meats. Buffalo rib-eye steaks, red deer (elk) chops, moose strip loin steaks, and red deer and sweet Italian buffalo sausages are among your choices. For an appetizer or the main meal, enjoy fresh smoked rainbow trout, individually vacuum-packed and ready to serve.

Or entertain in high-country style with one of the assortment packages that serve twenty as appetizers or eight to ten as entrées. The Krabloonik Applewood Smoked Sampler includes smoked trout, honey-cured smoked duck breast, smoked boneless pheasant thighs, and smoked red deer bratwrust. The Wild Game Sausage Extravaganza combines spicy Western buffalo and hot Italian red deer sausage with Muscovy duck bratwurst. I bet these tasty delights are not on your menu tonight.

MAURICE'S BARBEQUE　　　　　800-638-7423
P.O. Box 6847　　　　　　　　　803-791-8707 FAX
West Columbia, SC 29171

For a true Carolina treat, contact **Maurice's Barbeque** for gifts and home use.

Take a taste of the offerings with the Get Acquainted Special, which includes one rack of BBQ ribs, two pounds of BBQ meat, and two pints of Carolina hash. Or the Chopped Ham Barbeque sets your table with three pounds of meat perfect for bulky rolls. Either of these would make any group happy; with both you could throw a BBQ party.

When feeding a "small" extra-hungry crowd, of about twelve folks, the Barbeque & Ribs will eliminate the hunger pangs. Three racks of BBQ ribs and three pounds of BBQ meat will make a tasty buffet or ball-game pleaser. When creating your own favorites, you can use their sauces as a glaze or condiment to zip up pork, chicken, turkey, or beef. The regular, spicy, or hickory flavors could also easily be used with shrimp, trout, perch, or other fish.

NEW BRAUNFELS SMOKEHOUSE　　　800-537-6932
P.O. Box 311159
New Braunfels, TX 78131

For hickory-smoked excellence from the heart of Texas, **New Braunfels Smokehouse** delivers fine-quality smoked meats of this region.

Choose from the smoked turkeys or hams. Boneless or

not! All are ready for the buffet or family dinner table. For a special treat, there are golden smoked chicken and hand-rubbed smoked brisket. Smoked Pork Tenderloin, peppered beef jerky, or dried beef may be the very thing to spice up your party or outing.

When celebrating, try the Smokehouse Buffet! This sampler combines boneless breast of turkey, boneless peppered ham, smoked cheddar cheese, and a jar of sweet & spicy mustard. Or try the Party Makin's, including one pound each of smoked ham and turkey, one-half pound of sliced dried beef, a "chub" of salami, a ring of summer sausage, one pound of smoked cheddar cheese, and a jar of the sweet & spicy mustard! These hors d'oeuvres serve twenty or more people.

OMAHA STEAKS   800-228-9055
4400 South 96th Street   Discount: 10%
P.O. Box 3300
Omaha, NE 68103

Only in America's Midwest will you find cattle raised on the healthy diet of ripe golden corn. The corn is the secret to the tender and flavorful beef delivered by **Omaha Steaks**.

In their thirty-six-page catalog, you will find luscious cuts of Midwestern beef, pork, veal, and lamb. You'll also find seafood, tasty appetizers, and dreamy desserts. All the edibles are sure to be appreciated selections for any gift-giving occasion or the perfect ingredients for an unforgettable meal.

Try the Complete Delight, which includes filet mignons (one-quarter-inch cut), gourmet lobster tails (six ounces each), twenty-four assorted hors d'oeuvres, and Blank Forest cheesecake. Perhaps you have a more than hearty appetite and would enjoy a twenty-ounce T-bone steak or a thirty-ounce porterhouse steak. Hint: You may want to invite a friend or two over for this selection. Maybe chicken is on the menu. Experiment with the chicken Jamaica-style or breast of chicken cordon bleu.

OZARK MOUNTAIN SMOKEHOUSE 800-643-3437
P.O. Box 37   Discount: Cookbook
Farmington, AR 72730

For some of the best smoked meats to serve at a special occasion or tasty meal with plenty of leftovers, **Ozark Mountain Smokehouse** awaits your order.

All their hams are dry-rub sugar cured, which means there is no strong salty taste, unlike some that are cured with salt. Native hickory wood, which dries and preserves the ham while imparting a golden-brown color and wonderful flavor, is used for smoking. Upon arrival, the uncooked ham must be baked or fried before serving and will keep safely in the refrigerator for several months.

Ready to serve, fully cooked meats include hams, spiral-sliced hams, smoked turkeys, smoked chickens, and boneless ham and turkey breast, which are perfect for sandwiches or to adorn a buffet table. Slabs of bacon, peppered bacon, and smoked pork loin (also called Canadian-style bacon) are suitable with eggs Benedict and other brunch items, or even lunch or dinner.

Ozark Mountain Smokehouse also offers gift and party packs featuring their smoked delights along with smoked cheddar cheese and other tasty surprises.

Smoking Food

*T*he old-fashioned method of curing and smoking meats evolved
because there was no refrigeration. In the north, only the well-to-
do were able to stock their icehouses in the winter for refrigeration in
the warm months of summer. So if a person wanted to preserve a
slaughtered hog, he had to use the curing process that was naturally
available.

For this reason, animals were slaughtered in the late fall when
the temperatures were cold. The meat was then covered with the
"right amount" of cure—which consisted of salt, sugar, pepper, and
other spices—and wrapped in paper to keep the cure against the
meat. Then the meat was hung in feed sacks in a corncrib or smokehouse. The cold of winter prevented spoilage and allowed the cure to
penetrate the meat to the center.

For proper curing, the winter must not be so cold that the meat
freezes and stops the penetration of the cure (freezing otherwise does
not harm the meat). Too high a temperature would spoil the meat.
For this reason, keeping track of when to start the curing was imperative. (The states roughly along the Mason-Dixon Line have the best
year-round temperatures for successful natural curing, but the
method can be adapted north of this area.)

By spring, the curing was complete. The paper wrapping was
removed and the meat hung to dry. Over the following months, as
the hams, turkeys, pheasants, and other meats aged, they would lose
about 25 percent of their weight. The greatest gain in flavor occurs
during this period of hot, lazy days of summer and cool crisp nights.
This combination ages the pieces to their ultimate flavor. It takes
almost a year to reach this delicious point.

Today, there are still companies that cure and age their meats
naturally. They often use various curing processes employing different spices, salt substitutes for those with restricted diets, and a variety
of smoking materials. The curing/smoking technique has certainly

been refined to a science by now so that it produces consistent flavor that is unmatchable and a delicious treat.

Note that naturally smoked fish are prepared in much the same method. Instead of wrapping the fish, the pieces are generally smoked and dried quicker and don't need to age as long as meat. The old-fashioned method was as simple as salting the fillets, placing them on a rack about five feet off the ground, and allowing the sun to evaporate the moisture. A fire could be built under the rack if wanted, or spices could be added, but they were not always used.

PFAELZER BROTHERS   800-621-0226 Exc. IL
281 West 83d Street   312-325-9700 IL
Burr Ridge, IL 60521

Established in 1923, **Pfaelzer Brothers'** tender entrées have adorned tables at holiday feasts, backyard barbecues, and elegant dinner parties for over sixty-five years. Their filet mignons, boneless loin strips, and porterhouse and T-bone steaks are tasty cuts that will satisfy the most discriminating customer. All selections are available trimmed (side muscle and extra fat removed) or untrimmed, at your request.

Combination steak packages are not the Pfaelzer Brothers' only forte. Grace your table with beef Wellington or celebrate a promotion with panache—succulent lobster tails, salmon steaks, or gourmet stuffed chicken breasts. Hors d'oeuvres such as bite-size petite quiche or spicy stuffed jalapeños can kick off an elegant dinner party, or be a "four aces" choice for poker night!

A Food-of-the-Month plan is also offered. Send to friend or receive mouthwatering selections of in-season delights monthly, bimonthly, or on other convenient schedules. For instance, corn beef brisket and Irish whiskey cake make up the Saint Patrick's Day special. In addition, if you have been holding your breath for years for "chocolate-covered potato chips," Pfaelzer Brothers will be glad to send you a tin.

THE PUBLIC EYE   800-272 RIBS Exc. TN
301 S. Main Street   901-527-5757 TN
Memphis, TN 38103

When you need barbeque for a festive home event or larger gatherings, **The Public Eye** is ready to help.

Try the Rib Feast for Four when you have guests over. This delicious combination includes two large slabs of barbequed ribs, bottle BBQ sauce, and dry-rib seasoning. When you have a larger gathering order the Barbeque Pork Party, which is four pounds of chopped barbeque pork (enough for sixteen bulky sandwiches) and two bottles of barbeque sauce. Perfect for a buffet or a luncheon treat.

If you're undecided but swine dining is on the menu, try the Barbeque Combo Box. This all-inclusive feast contains one large slab of ribs, two pounds of chopped barbeque pork, a bottle of barbeque sauce, and dry-rib seasoning. This combination cannot miss with a hungry crowd.

RENDEZVOUS   800-524-5554 *BBCUE
52 S. Second Street   901-523-2746
Memphis, TN 38103

Do you want to have something in common with Frank Sinatra, Bill Cosby, and the Miami Dolphins? If so call **Rendezvous** for their famous barbecued ribs soon.

---

*Wait for the tone, then dial BBCUE (touch-tone phones only) or call the second number.

Whether you're feeding your family or putting on a feast, orders of these delicious ribs are certain to be a hit. The minimum order is five, which will more than feed five very hungry people. The delivery includes jars of their sumptuous sauce and seasoning, with heating instructions. After tasting the ribs and sauce, you are bound to want extra for your own cooking experiments. Bottles of special barbecue sauce and seasoning mix are available without the ribs.

SUGARDALE FOODS   800-275-1778
P.O. Box 571   216-832-7491 Collect
Massillon, OH 44648   Discount: 10%

For that special thank-you or congratulations, or even for your own table, **Sugardale Foods** will garnish the setting graciously.

A panoply of ham flavors is just part of their offerings. Try the double-smoked Beauty Ham, whose rich flavor and tenderness are achieved in two leisurely visits to the smokehouse. Before the second stopover, the ham is lightly scored and glazed with brown sugar, cinnamon, and a blend of spices that penetrate the meat. Other palate-pleasing ham flavors include honey-cured, peppered, honey-mustard, Cajun, and Signature ham, which is traditionally smoked and ready for your favorite glaze.

Winning combinations of steaks and roasts for any occasion will make for a cherished meal. Choose from the convenient packs sized for any event. Boneless strip, porterhouse, and rib eye steaks are among the selections. Filet mignons, boneless prime rib, or chateaubriand are bound to make your meal or gift a memorable feast.

Sugardale Foods has many other delicious, ready-to-heat and -eat meats. Gift baskets and four-, six-, and twelve-month meat-of-the-month selections are also available.

Chapter 8

Coffee, Tea, Wine and Spirits

Making life an adventure rather than a routine can start with the morning coffee or tea. Yet many people do not explore the vast, almost endless varieties available. After breakfast the adventure can continue through the day and culminate with a special evening brew. After all, we would not normally consider having bacon and eggs for breakfast, lunch, and supper. Then why limit our coffee or tea options?

Standard all-American ground coffee, which has slightly different tastes, but is not significantly unique, is the common choice, with the only major variant being how much milk or sweetener you add. New versions of tea are more popular now, but the options are generally limited. This chapter will give you many new and delicious ways to enjoy the two old favorites, including numerous new flavors.

Choosing the appropriate wine for a meal involves a roll-of-the-dice decision for many people. The general guideline for most wine drinkers, and probably learned in college by some self-proclaimed "expert," is to choose red wines for red meat (or anything red like spaghetti sauce or chili), and white wines for "whiter" meats like chicken or fish. This works most of the time, kind of, but there are many dishes that are exceptions. A dry red Burgundy would not complement salmon, for instance. And what goes with the fresh fruit served for dessert?

Wine makers go to great lengths to produce a certain taste and bouquet (aroma). To pick the correct wine or wines to serve is really a matter of your taste, but with a few simple tips and some experience, the not-so-subtle differences will be quite evident. I highly recommend a wine-tasting course. They are generally not very expensive and will enlighten you on the best wine suitable for a particular food.

For those of you who aren't novices in choosing wine, the general "rule-of-palate" is to pick a wine that's acidity or sweetness is more than that of the food being served. If a salad with an acidic dressing is part of the meal, be certain the wine is more acidic and the tastes will complement each other. When the wine is suitably matched to the food it's being served with, you can present a five-dollar bottle of Beaujolais and pass it off as a premium fifteen-dollar wine. For desserts, always find a sweeter wine than the dessert you are serving. A French Sauternes is perfect with fruit, but will not work well with cheesecake. Try a German Berre Alsace with the sweeter dessert.

There are many books to guide you to the appropriate wine for the food, but only experimentation can teach you about choosing the correct wine for the food; and remember, you don't have to spend a fortune for appropriate wines. For more information on reasonably prices wines, write: The Wine Guide by Stacy Goss, Guide Associates, Inc., 98 Wadsworth St., Suite 127-228, Lakewood, CO 80226. Include a six-dollar check, which covers the cost of the guide and shipping.

BROWN & JENKINS TRADING CO.   800-456-5282
P.O. Box 1570
Burlington, VT 05402

You are invited to join the **Brown & Jenkins** coffee club. Warning: The club is only for true coffee lovers!

Delightful, mild, smooth, medium, "secret," classic, and unique blends are only part of the sampling of Brown & Jenkins coffees. For the true connoisseur, each private gourmet house blend is developed to satisfy a morning, noon, or night occasion.

The over a dozen Selected Choice Coffees along with the Private Gourmet Blends and Special Roasts, totaling over thirty, will accommodate both pedestrian and gourmet palates. For instance, flavored coffees like amaretto and Viennese cinnamon are blended especially for after-dinner enjoyment.

Decaffeinated coffees are not forgotten! The flavored selection, European process, and Swiss water process blends are all available in many varieties. Put your cup or mug down and call for a Brown & Jenkins brew. Delivery schedules can be designed (always including a free sample) so that you will never run short.

COMMUNITY KITCHENS   800-535-9901
P.O. Box 2311, Dept. GN   800-321-7539 FAX
Baton Rouge, LA 70821

For a wide variety of food items for gifts or personal use, **Community Kitchens** has something for everyone.

This company has a unique selection of coffees as well as practically any other food item or kitchen ware. Their Private Reserve Coffees are superior to many others generally available. Two of the choices are the Evangeline Blend (named after the heroine in Longfellow's poem), which is a dark coffee combining a spectrum of flavors—from the most winey, like Kenya AA, to the driest, like Santos Bourbon. The result is a coffee connoisseur's dream. Plantation Blend is another exceptional brew which combines African, Brazilian, and Central American beans, resulting in a perfectly smooth, full-bodied taste.

A large collection of Louisiana coffees is also available. As you may expect in the home of one of America's most distinguished regional cuisines, they do not drink ordinary coffee. In addition to the various brews they sell, an extensive list of many types and styles of foods are sold, from pasta to marinated vegetables and more.

Brewing the Perfect Coffee

*U*ndoubtedly, you have had bad coffee. Either weak, burnt, or somebody washed the pot but forgot to rinse it. Perhaps you can depend on some coffee to be good. But have you ever really found the perfect brew? If you want to explore the wide world of coffees for that perfect cupful, start by going to the source—coffee beans.

Experiment with different coffee beans to identify their distinctive flavors. Making notes as you go is probably a good idea. Indonesian beans are spicy, earthy; Kenya AA has a winy taste; Kona is light-bodied, unlike Colombia, which has a full body. After appreciating the varying tastes of these and others, start combining them in search of the perfect brew.

Storage of the beans to maximize their freshness is imperative. Keep them in an airtight container in the refrigerator if they are to be used within a week, otherwise freeze them but be certain to bring them to room temperature before grinding.

Always grind the beans just before brewing. If your coffee is bitter and overly strong, the grind is to fine or too much coffee was used. When the coffee lacks flavor or strength, the grind may be too coarse or more coffee should be added. And please, if the coffee is too bitter, do not add salt to "mellow" it. You simple end up with salty, bitter coffee that could probably take varnish off a table.

The "rule-of-brewing" is to use two tablespoons of freshly ground coffee for every 6-ounce serving. Since coffeepots are seldom marked in six-ounce increments, but in cup numbers, the amount of ground beans will have to be adjusted accordingly.

As you are closing in on the perfect brew, you may want to switch to spring water for the fullest flavor. Softened and distilled water will distort the characteristic of the coffee blend. Softened water has salt in it. Distilled water lacks minerals, resulting in a flat taste. Be sure to heat your water to almost boiling to extract the optimal flavor.

All coffee, even the burnt variety, is always better fresh. When you only need one or two cups, only make that amount. After going through the trials and tribulations of discovering the perfect brew, don't let it heat and reheat on your drip machine. After ten or fifteen minutes, it will taste awful. Pour the fresh coffee into a glass-lined Thermos, which will keep it warm until the next cup is needed. If it cools off, "zap" it with the microwave, and the fresh-brewed flavor will still be there.

 GEORGETOWN COFFEE, TEA & SPICE
1330 Wisconsin Ave., N.W.   202-338-3801
Washington, DC 20007

If your first wakeup call is the sound of a coffee grinder, **Georgetown Coffee, Tea & Spice** is waiting to wake up your palate.

Try a couple of the over thirty varieties of coffee from around the globe and you will be hooked. Instant will just not start the morning anymore. Choose the African Blend for a full-bodied flavor when your work schedule calls. But enjoy the Colombian Supremo when a mild flavor and smooth aroma are the company you need with slippers and the Sunday paper. All the varieties come in bean form or your choice of grind.

Specialty coffees, unroasted green beans, and exciting flavored coffees are also available. Caffeinated, Decaf, and after-dinner samplers are perfect gifts or introductions to the world of coffee. Don't miss the large variety of teas, regular and decaf, spices, and their famous Italian salad dressing mix.

GEVALIA KAFFE
P.O. Box 11424
Des Moines, IA 50336

800-678-2682
515-284-3775 FAX

If you are getting tired of the same old coffee in the morning, maybe it is time to try some tastes of Sweden from **Gevalia Kaffe**.

Compare three of the varieties of blended coffees. Traditional Roast is smooth and velvety—with high flavor notes and no bitter aftertaste. Dark Roast is bolder, stronger, deeper—with the pleasant bittersweet undertones that are attractive to dark roast aficionados. Espresso is the darkest roast, with a rich, pungent, deep flavor. The sought-after bitter notes are retained without the muddy, burnt taste found in many high-roasted coffees.

If you prefer a flavor in your coffee, choose from the assortment of flavored coffees. These include: cinnamon, amaretto, mocha, Irish creme, vanilla nut, and hazelnut. All are ideal for after dinner, with perhaps a touch of your favorite liqueur.

All the blends and flavored coffees offered by Gevalia

Kaffe come in caffeinated and naturally decaffeinated varieties. Their product list also contains many edibles to complement the coffees.

LIQUOR BY WIRE   800-621-5150
5252 North Broadway   312-334-0077 IL
Chicago, IL 60640   312-334-2438 (24 hr.)

For delivery of virtually any spirit, wine, or beer the company you want to call is **Liquor by Wire**.

For gifts to a colleague or to say "thank you" for a job well done, all are available for your gift giving. Pick from the extensive catalog to find that special libation for the occasion. The enclosed toast and gift card will only enhance the pleasure of the folks receiving favor.

Try the specialties section, which includes an imported beer assortment with 10 varieties in a colorful gift case...a five-pound Hershey's milk chocolate bar...perhaps the Wine-Bo Bottle Bow, holding one 750-ml-size bottle and assorted "goodies."

ORLEANS COFFEE EXCHANGE   800-537-2783
Discount Coffee Outlet   800-743-5711 FAX
712 Orleans
New Orleans, LA 70116

With prices such that many can now afford the finest coffees, the time has come to explore **Orleans Coffee Exchange**'s vast selection.

Choose from the over two hundred flavored coffees, with palate pleasers like Black Forest cake, Cinnamon Apple, Mint Julep, or Super Vanilla. If a little extra boost is needed in the morning, all come in the "Wake Up" variety, which has

35 percent more caffeine. Or try one of the more than forty house blends, decaffeinated, Swiss water decaf, expresso, or a grind from the assortment of Natural Beans.

Perhaps teas are your fancy. The international flavors from Ceylon, China, Taiwan, Japan, and India can adorn your table with only a phone call. Dozens of flavored and special teas are available in regular and decaffeinated styles. Don't overlook the instant coffee beverages, expresso, and bulk packs.

SHERRY-LEHMANN 212-838-7500
679 Madison Avenue 212-838-9285
New York, NY 10021

GEORGES DUBŒUF

CHIROUBLES

APPELLATION CHIROUBLES CONTROLÉE
RED BURGUNDY WINE
MIS EN BOUTEILLES PAR
LES VINS GEORGES DUBŒUF
71720 ROMANÈCHE-THORINS

PRODUCED AND BOTTLED IN FRANCE

ALC. 13% BY VOL.   CONTAINS SULFITES   750 ML
IMPORTED BY: W.J. DEUTSCH CO. CHAPPAQUA, N.Y.
SPECIALLY SELECTED FOR SHERRY LEHMANN INC. NY, NY

To choose from the largest collection of the world's great wines is a privilege. **Sherry-Lehmann** extends that courtesy to serious wine collectors and gift givers.

The hundreds of selections offered are a veritable "candy store" to collectors and wine or champagne aficionados. Regions, such as Bordeaux and Burgundy, and countries are featured, with vast selections and each wine's qualities described. The vintages of each country are also described, and the best of a nation's vintage is offered by the bottle or the case.

Investors can buy "futures" in yet to be bottled wines. The eighties produced seven banner years of award-winning wines, which doubled an investor's original investment, or better. You do not have to be wealthy to buy "futures" and they make a fantastic gift for those who appreciate fine wines.

Home Wine Making

*W*ine making may create images of barefoot women stomping rhythmically in a large vat, with a backdrop of a rolling vineyard. Perhaps you have toured a stainless steel complex that churns out the libation that makes lovers swoon in each other's arms. Home wine making is somewhere in between.

If you have ever toyed with the idea of making wine, or another brew, delay no longer. I have found the hobby fascinating. My mentor taught me to change five pounds of rhubarb into five gallons of "chablis/blush-like" wine! Over the years, I have produced over one hundred gallons of wine for gifts, parties, and personal consumption. To this day, I find the hobby engaging as I search for the perfect combination of ingredients to produce a prize-winning wine.

The first step in the process is to mix a combination of juice, yeast, and nutrients to create the "must." Juices are extracted from grapes, other fruits or berries, flowers, and even vegetables! (Note: non-poisonous ingredients are preferable if you plan to serve the wine to friends!) The goal is to extract the taste and bouquet (the aroma) from the edible. For the beginner, buying a concentrated juice at a wine store is the easiest way to get acquainted with the fermentation process.

Within days, the "must" mixture will start foaming and bubbling. The reaction of the natural sugars and yeast is producing carbon dioxide at this point. As the days pass, solid matter, such as fruit skins, seeds, dead yeast, etc., drop to the bottom of the fermentation container.

Before the science of wine making was perfected, home brewers, perhaps your grandfather, simple crushed grapes or apples and put them in a wooden barrel. The two essential catalysts to the fermentation process—yeast and nutrients—exist in nature. Yeast is in the air we breath and the rudimentary nutrients are contained in the skins or solid matter that you are making the wine from. So after a few months, grandpa had produced a libation with a similar alcohol content as present-day drinkables, but not as refined of impurities, hence he would need aspirin the next morning.

When producing wine at home, siphoning the clear liquid refines the taste and stabilizes the bouquet. With the primary fermentation of "must," the liquid (wine-to-be) is siphoned into a five-gallon glass container after about ten days, leaving the sediment, or dregs, behind. This process is known as "racking."

At this point the storage container, or "carboy," is kept in a cool area—no warmer than 65 degrees. Most basements are adequate for this part of the fermentation process. Additionally, an airlock is installed so contamination from fruit flies and other sources will not change your wine-to-be into vinegar.

The racking procedure is repeated every three or four months when about one half inch of dregs has settled to the bottom of the carboy. The wine will become clearer between each racking. Yet patience is imperative. The quality will only improve if the wine is aged longer in the carboy before bottling. The fermentation will stop naturally as the alcohol content kills the yeast—about 12.5 percent is the upper limit. At this point, no additional dregs will settle and the wine is ready to bottle.

All hobbies cost money. A beginner wine making set, juice concentrate, and all the needed equipment can be purchased for about $70. Each additional concentrate or fresh squeezed juice can be *bought for $18-$25. Each concentrate will yield about five gallons, or about twenty-eight bottles of wine. There are many recipe books available if you are considering using your own fruit, flowers, or other fermentables. An additional investment of $50 to $200 will be needed in this case for a press and other equipment to balance the taste of the wine.*

So you need not kick off your shoes and stain your feet, or buy expensive equipment to become a wine maker. The results will be surprisingly pleasant when you pop the first cork with family or friends.

STASH TEA	800-826-4218
9040 S.W. Burnham Street	503-684-7944
Tigard, OR 97223	503-684-4424
	Discount: Coupon
	for assorted teas in mug

If you haven't tried loose teas, you'll be surprised by their ease and convenience. Let **Stash Tea** introduce you to the world of loose tea.

The loose teas offered are divided into four categories. Spiced tea blends include such delicate to zesty blends as almond cream and jasmine spice to lemon and orange spice. The spiced herbal blends, such as apple spice or licorice spice, are ideal with dessert or as afternoon pick-me-ups.

From the classic herbal teas section you will find soothing flavors like peppermint/lemongrass, sandman, and ruby mist, which is a combination of hibiscus, rosehips, mint, and a tart citrus blend. Then there are a traditional tea blends: darjeeling, Earl Grey, and orange pekoe, to name a few.

Most of the loose teas are available in either foil-sealed teabags or paper-wrapped teabags. Stash Tea's catalog also offers accessories, sampler packs, and gift packs for that tea connoisseur in your life. Break out of that same-old-tea rut, try some citrus spice or wintermint soon.

Brewing Hints for Tea

*T*he teabag is a relatively recent invention. For centuries, people enjoyed their favorite beverage without teabags—and made brewing part of their pleasure.

Today, there are many ways of making tea. All involve the same four "ingredients": water, tea, a straining device and a cup. Making tea is a simple process; after all, you are just steeping leaves in hot water until they release their flavor. But everyone has their favorite method.

If you use teabags regularly, you have made the process one step quicker. To the chagrin of tea lovers, I use a teabag, placed in a mug of cold water, and "zap" it with the microwave. But there is certainly a vast difference between my tea and the properly brewed version. Here are a few hints:

● *To brew for the best flavor, start with fresh, cool water from the tap or bottle. Never use water from the hot tap.*

● *Pre-warm your pot or cup. A cold pot or cup will cause the water to drop below perfect brewing temperature.*

● *Pour water at the first rolling boil. Aerated water brings out the flavor more. Overboiled or standing water gives tea a flat taste.*

● *Brew four to five minutes to develop full flavor. Brewing time depends on the blend and your personal preferences. Herbal teas take a bit longer than true tea.*

● *Remove the tea when brewing is complete. Don't squeeze the teabag. Wringing or overbrewing brings out tannin in true tea and creates a bitter taste.*

Most teabags are measured to make one large mug or two five-ounce cups. Use the teabag only once; for two cups or more use a pot for consistent flavor.

Using loose tea can be a new experience for many people. There are four methods to create the afternoon beverage. With a tea press, you drop your favorite loose tea in the coffeepotlike container, pour on the water, and watch it brew. When it is the right strength, you press down the filter, trapping the leaves on the bottom. Or for the ultimate ease, an automatic tea maker does it all, from boiling the water to timing the brew to suit your taste.

If you prefer straining tea the old-fashioned way, or just want to give an occasion a touch of elegance, nineteenth-century designs that fit over a mug to produce single cups are available. This also allows you to serve a number of flavors when you have a gathering. Infusers do equally well with loose tea. Their clawlike round cage holds the tea so that it may be dipped into the hot water easily, and removed when the brew is perfect.

For loose tea, a teaspoon per cup is an American norm. You may prefer a pinch, as in the Orient, or a strong concentrate, like the Russians. Whatever your choice, enjoy experimenting with the dozens of flavors offered by the companies in Feast.

800 SPIRITS INC.   800-238-4373
Two University Plaza   201-342-8269 FAX
Hackensack, NJ 07601

If you need a special libation to celebrate closing a deal, congratulations for a job well done, or glad tidings, choose from the extensive selection of spirits from **800 Spirits Inc.**

Most any product available in your local liquor or wine store is available through the mail. This company is a worldwide gift delivery service for champagnes, wines, spirits, and a wide assortment of complementary accessories. With your order, cakes, chocolates, ad even gourmet pasta (Rossi Pasta, chapter 4) can be included to accent the gift.

The offerings include: aperitifs, American, worldwide, and kosher wines, whiskeys, gin, vodka, brandy, and more. The drinkables featured are generally medium to better in quality and are handsomely gift wrapped according to the destination. If you are sending a bottle of Suntory Royal Scotch to a friend in Japan, it will be wrapped in the traditional Japanese fashion with your gift message written in Japanese.

Chapter 9

Condiments, Spices and Herbs

If you enjoy hearing "What did you put in this to give it that wonderful taste?" from your dinner guests, this chapter will give you practically endless "secret ingredients."

Condiments, spices, and herbs add a new dimension to the simplest of dishes. If you have dietary restrictions on using sugar, or you are trying to cut back, there are many syrups and honeys that will achieve the same sweetness but are healthier. The substitutes also add a wonderful new taste to the same old, but good, dish. Note later in this chapter how to replace sugar with substitutes in cooking.

Spices and herbs must always be respected. When cooking, add them slowly for perfectly balanced taste. If you add too much, the spice will not be a complement, but the overriding attention grabber of the dish. In that case, you may have to start over, or as I do, double all the regular ingredients to "water down" the spiciness. I ended up with about five gallons chili once due to my lack of finesse with hot peppers.

Always note the storage directions. Herbs and spices

should be kept in a tightly closed container, away from heat and light. They should also be replaced after a year, or use more than the recipe calls for to produce the same spicy effect. Condiments often need refrigeration to keep from spoiling, particularly those without preservatives. So always read and heed the warnings on the labels for the best results.

GLORYBEE NATURAL SWEETENERS, INC.
120 N. Seneca Road   503-689-0913
Eugene, OR 97402   503-689-9692 FAX

If you are looking for a wide selection of honey, honey prod-ucts, and natural sweeteners, **Glorybee Natural Sweeteners, Inc.**'s expanding product line is sure to interest you.

For 100 percent natural, raw, uncooked, certified, (kosher) honey, choose from the fourteen sweet varieties to add to your favorite recipe, stir in tea, or use as a substitute sweetener for refined sugar or artificial alternatives. Clover honey, usually what you buy in the grocery store, is available, but try some raspberry, orange, bandon cranberry, or huck-leberry as the secret ingredient in your next batch of muffins or in a cake.

Their newest product is "The Candy of the Future," Honey Stixs. The candy is stick-formed honey flavored with pure extracts. Some of the seven flavors include lemon, cool mint, lime, and ginseng. They can be eaten alone, used to stir a hot drink to give it a special taste, or broken up and sprinkled over desserts.

GOURMET TEMPTATIONS INC.
1111 E. Francisco Blvd., #4
San Rafael, CA 94901

415-456-1332
415-459-7671 FAX

Imagine pouring whole boy-senberries or raspberries over your morning pancakes or waffles—that's the Fruit Indulgence offered by **Gourmet Temptations Inc.**

The luscious Fruit Indulgence produce line of fruit toppings/spreads are fruity delights to be enjoyed with every meal. Spread the treats on breakfast breads, atop ice cream, yogurt, cheesecake, and other desserts for a healthy, natural alternative to sugar syrups containing only a hint of real fruit. Bake with your newest ingredient—use as a filling for cookies, cakes, pies, or as a sugar substitute in many of your favorite recipes.

The flavors offered will match almost any cooking or eating need and include marion blackberry, raspberry, boy-senberry, strawberry, blueberry, bing cherry, mixed berry, and peach. Try one of the flavors in your finest marinades and sauces to bring out a new character and taste experience. The Peach Indulgence would be perfect for a glaze on chicken or game, turning an ordinary meal into something special.

Beekeeping

*H*oney—the sweet syrupy substance that coats sore throats—is unmatchable over scones, and is a pleasant substitute for processed sugar in most recipes. Around 400 B.C., the Egyptians fed the sweet nectar to sacred animals before they were sacrificed to the gods. To the surprise of archaeologists, Egyptian tombs sometimes included containers of honey. In eleventh-century Germany, sticky jars even passed in the marketplace as currency!

Through the ages, the sweet nectar from the buzzing hives has been harvested and revered as a special edible. The supplier, the bee colony, remains a well-organized society, with members having distinct purposes important to the survival of the group.

The colony consists of a queen, drones (male), and workers (female). A single strong hive will harbor over 60,000 bees. Only 1 or 2 percent of these are drones!

The Queen's function is to lay the eggs, sometimes up to 1,800 a day! She is born an ordinary worker until the other workers feed *her a rich mixture of food called "royal jelly." After feasting for 16 days, she develops the ability to lay eggs and hence is recognized by all the hive members as the queen.*

The drones also have a single purpose if life. Their labor is to fertilize the queen. When the queen has laid the maximum number of eggs for the colony, the drones' lives end. They are forced out of the hive by the workers. The drones are the expendable members of the society and are driven out of the colony as winter approaches. They eventually die from cold or starvation.

The workers, as their name implies, work. Their jobs include gathering nectar and pollen, feeding young larvae, and warming/cooling and protecting the eggs, larvae, and pupae. In addition, the workers supply water and build the combs to hold the honey. In essence, the workers oversee the colony and every aspect of its operation. They are also the only bees that have "stingers," which they use aggressively to defend the hive.

Commercial honey producers arrange hives in stackable "supers," which are similar to bureau draws. The lowest super is the brood chamber. This is where the queen resides, the young honeybees are raised, and extra food is stored for the winter months. The upper supers are where the honey is produced in the honeycomb. These sheets of sweet substance are harvested twice each year—once in the summer, and again in the fall.

Many mail-order companies that specialize in honey produce their own special varieties. Others stock over a hundred American varieties. Silver City Apiaries, Peterborough, New Hampshire, maintains about 300 colonies scattered across southern New Hampshire. Each colony produces between thirty and one hundred pounds of honey each year. Much of the discrepancy in yield depends on weather, as the flowers or fruit trees whose nectar is gathered and turned into honey may or may not be "fertile" due to a dry spell or a sudden cold snap.

Most honeys vary in flavor due to weather conditions that affect the host flowers' "fruit." A very dry year may produce a "weedy" small plant or flower. Yet a wet growing season produces a robust healthy bloom. The amount of honey each colony produces is equally dependent on the season.

Many grocery stores have a limited supply of honey—generally clover. Since the taste of the product is derived from the type of flower, experimenting with, say, orange blossom honey in your favorite recipe may produce a new and better taste.

Explore the companies listed in Feast. *If your preference is regional or a refined culinary statement is needed, experiment with varieties other than clover.*

GREEN BRIAR JAM KITCHEN   508-888-6870
6 Discovery Hill Road
East Sandwich, MA 02537

When you are looking for characteristic condiments that say New England, **Green Briar Jam Kitchen** has a fine collection to choose from.

Your morning toast or bagel could only be better with an assortment of jams, jellies, and marmalades to choose from. Pick the apricot or ginger rhubarb jam or maybe apple rum jelly, perhaps followed by lemon-lime marmalade. Ginger orange jam or quince jelly can be used to baste your poultry dish.

Pickles and relishes are certainly not forgotten. The classic bread and butter variety is available, but if your tastes are more daring try pumpkin pickle, hot pepper relish, or Martha's mustard. Sun-cooked fruits and specialty items, such as yellow tomato with ginger, are also offered.

The list from Green Briar Jam Kitchen includes gift assortments and samplers that will only make you want more.

INDIANA BOTANIC GARDENS   219-947-4040
P.O. Box 5   Discount: 10%
Hammond, IN 46325

Looking for a company to supply natural products to improve your own and your family's health and well-being? Since 1910, **Indiana Botanic Gardens** has been doing just that.

Over 1000 different herbs, herbal blends, teas, and other natural products prepared from all-natural ingredients are available. In the extensive catalog, the products are explained as to their medicinal and health properties. The medicinal offerings range from laxatives, kidney and bladder problems, high blood pressure, to simple pick-me-ups if you are feeling tired.

Spices, seasonings, and liquid herb extracts are also sold. All are said to be health oriented. Within the assortment of herbs, you will find every type, from the common, such as horehound or spearmint, to obscure roots, barks, and seeds. Their uses are explained and how to prepare them for consumption is detailed.

JASMINE & BREAD   802-763-7115
RR# 2, Box 256
South Royalton, VT 05068

If you find that the catsup, mustard, and barbecue sauce you buy aren't the best, and could easily be done without, you are definitely not purchasing **Jasmine & Bread** products.

Beyond Catsup is an exciting blend of tomatoes, apples, vegetables, cider vinegar, and spices. It can be used in place of catsup, but also is great as a basting sauce, in marinades, sauces, salad dressings, and dips. Plum Perfect is a sassy sweet & sour sauce made of plums, apples, cider vinegar, honey,

---

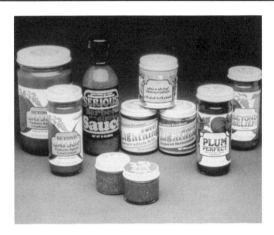

mint, and spices. No salt! This product will enhance the flavor of lamb, smoked turkey, chicken, pork, and wild game. It is also an excellent barbecue sauce or dipping sauce for Oriental dishes, hors d'oeuvres, and more.

Other selections include: Beyond Belief, Beyond Horseradish Mustard, White Lightening, Sweet Lightening Horseradish Jelly, and Serious Barbecue and Dipping Sauce.

KIMBERELY WINE VINEGAR WORKS   415-755-0306
290 Pierce Street
Daly City, CA 94015

In the world of wine vinegars, using oak barrels and ignoring accelerated production are rarely the case, but the exception. **Kimberely Wine Vinegar Works** is the exception.

Most vinegar starts as wine diluted by one third with water. These vinegars are undiluted and meet the strength of vinegar condiments that is expected from fine cooks and those of use who just want to add pizzazz.

The zesty treats include cabernet sauvignon, sherry, and garlic cabernet, all wonderful salad and cooking additions.

Extra-virgin olive oil and chardonnay wine vinegar provide the addition of perfect "secret ingredients." For the cook in the family, the selection is a perfect gift.

KITCHEN SECRETS   800-527-1311
P.O. Box 11147   214-324-2434
Dallas, TX 75223

If you have ever wanted to make your own sausages, and needed the right spice mix and expert recipes, **Kitchen Secrets** is ready to help.

The blends come in six variations. All the sausage seasonings make a great treat when you are eating at home or bring a special hors d'oeuvre to a party. Each is easy to make and customers' recipes are supplied, but you are encouraged to try your own, also.

The seasoning blends are summer sausage, pepperoni, salami, pastrami, bologna, and chili. Beef, turkey, pork, and similar meats can become your signature sausages. For the hunter of the family, venison sausages will be a special treat or a perfect addition to a birthday dinner.

A sampler pack and a hunter's kit are part of the offerings by Kitchen Secrets. So call soon for free recipes and a guide to making our own sausages.

MANSMITH ENTERPRISES, INC. 800-627-4981
Adjacent to the San Andreas Fault 408-623-4981
600 Mission Vineyard Road 408-623-2150 FAX
P.O. Box 376
San Juan Bautista, CA 95045

Are you looking to experience the cutting edge of the BBQ industry? Contact **Mansmith Enterprises, Inc.**

Their trend-setting product line is deliciously different, fun to use, and brings out the creativity of the backyard chef. Their grilling spice is too "grate" for words. The charismatic blend of select herbs and spices contains *no sugar* and is formulated to enhance the flavor of almost any food. If your special creation is soup, salad, vegetables, casseroles, or dressings, the convenient seasoning is a must for that perfect taste.

Try the only BBQ Paste on the market. When firing up the grill, you will have over a hundred barbecue sauce flavors for pork, poultry, seafood, beef, lamb, and all the smorgasbord of items we enjoy grilling. Making your own taste of BBQ sauce is easy. Just mix the BBQ Paste with an equal amount of such liquids as water, beer, wine, fruit juice, soda, hard liquor, or whatever you prefer.

MEADOWBROOK　　　　800-845-0008 Exc. CT
Herb Garden Catalog　　　203-254-7323 CT
P.O. Box 578　　　　　　　203-254-3571 FAX
Fairfield, CT 06430

Rediscover the highest-quality herbs straight from a New England garden into your favorite recipe, supplied by **Meadowbrook**.

Choose from the over seventy herbs and spices offered, to fill the empty slots on your spice rack. The individual spices are perfect to construct a new-cook spice gift kit for the newly married or first-time apartment dweller. For the easy-to-use blends, pick from the six mixtures: fish herbs, hamburger seasoning, poultry seasoning, salad herbs, heart's delight, and soup herbs.

Change the same old meal, even though it may be good, into special fare with Timesaving Herbal Recipe Blends. There are twelve delicious blends offered, including ultimate garlic bread, party dip herbs, potato & herbal salad dressing, boursin-style cheese spread, and more. Each recipe blend has two or more recipe variations that will suit your culinary needs.

Don't miss the many other products that Meadowbrook has, including mustards, herb and fruit vinegars, and health aids.

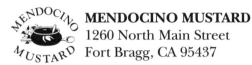

MENDOCINO MUSTARD   707-964-2250
1260 North Main Street
Fort Bragg, CA 95437

For a taste of the finest, call **Mendocino Mustard** for their nationally known condiment which graces the kitchens of elite restaurants and food shelves in the country.

This zesty, no-sodium mustard has long been a favorite on sandwiches or in special recipes. Mix it with your favorite glaze for meat or even whisk a dab into scrambled eggs to liven the morning's breakfast. With either choice, you will not be disappointed.

This hot sweet mustard is the company's only mustard, but as Devora Rossman says, "We produce only one product very well." They pride themselves on creating the perfect mustard and not contracting to a commercial facility. The TLC shows.

PENDERY'S   800-533-1870
304 East Belknap Street
Ft. Worth, TX 76102

If your impression of Texas cooking involves too many hot peppers, the selection at **Pendery's**, established 1870, will soon change your mind. Their catalog cover proudly announces—"Welcome to the World of Spices."

For the Southwestern flair, choose from the chile pod wreath or perhaps a jar of mole, which is a paste containing chiles, sesame oil, sesame seed, peanuts, cinnamon, coca, and spices. Or try the popular Red Hot West or Western Bandana Basket, which is equally appropriate for the casual cook or gourmet chef.

Maybe you would like to single out a few hard-to-find spices or herbs. Some examples of the over fifty that are offered include star anise, calendula, epazote, fenugreek, ginseng, Mexican oregano, and seaweed. Some more conventional selections are offered: turmeric, white pepper, and mustard seed.

Blends and seasonings from Cajun-style to chicken seasonings are just part of their selection. Harder to find Garam Masala is an Indian seasoning used in curries. Other selections include mixed pickling spice to add to stews, soups, relishes, and vegetables as well as in pickling and preserving meats. Many gift mixtures are available.

Even herbal teas, honey, fruits, and candy are part of the diverse offerings from Pendery's. Call today for those hot chile items, but do not overlook the other wonderful tastes from the Southwest, or you will be missing out.

PICKITY PLACE   603-878-1151
Nutting Hill Road
Mason, NH 03048

Enjoy the exquisite tastes from a two hundred-year-old home, where gardening and herb harvesting have become a way of life, **Pickity Place**.

Choose from the many gifts, baskets, or hard-to-find condiments, herbs, and spices. Their herb tins make great hostess or shower gifts. A sampling of the selection includes fresh basil, chives, nutmeg (ground), rosemary, and many more. Maybe a selection of mustards and flavor enhancers fits the bill. Crushed garlic in water and the favorite basil sauce, pesto, are the perfect additions to any cook's cupboard.

Maybe some of the same herb blends that are used in their famous restaurant could enhance your favorite recipe. Salad herbs add a wonderful summertime taste to salads year-round. Mulled cider mix is a unique blend that can be added to apple cider, juice, or wine. Herb cheese mix is a tasty blend of garlic, sea salt, and spices to add to yogurt, cream cheese, or sour cream. Many more blends are available.

Garnishing With Flowers and Herbs

*C*ooking with flowers and herbs has long been popular for adding that special aroma or splash of color to many recipes. Oftentimes, flowers can even be used as the main ingredient. The results are frequently first-rate as the subtle flavors and textures add that intimate taste and create a dish that stands out.

Both the leaves and the flowers can be used on some plants. Hyssop, thyme, marjoram, and rosemary are some examples. Nasturtiums and marigolds are perfect in salads, adding a delicious, spicy flavor and brilliant color. Rose petals are not only an attractive garnish, but are used in summer puddings and produce a discrete, attractive aroma. Elder blossoms can be put in summertime drinks. Roses and lemon-scented geraniums make delicious preserves, while dried flowers can be used to make fragrant teas (note the Stash Tea Co. in chapter 9).

When flowers are used in recipes, they should always be fresh. Never use any that have been sprayed. Check carefully for insects and wash the flower gently since the petals can be bruised easily. Please also note that some flowers or their leaves are poisonous, so checking the many published guides is wise before eating them.

For Further Information and Recipes:

Newdick, Jane, & Lawrence, Mary. The Miniature Book of Flowers as Food: *New York: Outlet Book Company, Inc., 1991.*

ROWENA'S
758 West 22d Street
Norfolk, VA 23517

800-627-8699
804-627-8699
804-627-1505 FAX

The quiet, nocturnal cooking sprees of a woman expecting her first child gave birth to this thriving food business: **Rowena's**.

Try some of her "curd" flavors. This custardlike sauce makes a perfect frosting or ingredient. This traditional English delicacy is offered in lemon, raspberry, and pineapple. Perfect on cakes, scrumptious topping ice cream, or as a filler in cream puffs or eclairs. Raspberry curd is especially good on Dark and Delicious Chocolate Pound Cake or your special brownies.

For the youngsters, *The Adventures of Rowena and the Wonderful Jam and Jelly Factory* is a delightful children's story-cookbook which will enchant children and encourage their creative skills while preparing easy and delicious recipes. Great for our young beginners.

Rowena's also sells sauces, peanuts, baking mixes, and gift packs.

SAN FRANCISCO HERB CO.   800-227-4530
250 14th Street   800-622-0768 Canada
San Francisco, CA 94103   415-861-7174

If you are looking for a wide selection of the most commonly used spice blends, herbs & spices, and dehydrated veggies, take note of the **San Francisco Herb Co.**

Within their Spice Blends section, they present fifteen blends, including barbecue, Cajun, chili, lemon pepper (perfect for fish dishes), pickling spice blend, and more. All come in either one- or five-pound packages and are in "ground" or "whole" form.

Offered in the Culinary Herbs and Spices area are over one hundred selections. From allspice to fennel seed to juniper berries and vanilla beans, in one- and five-pound packages, and a few conveniently furnished in four-ounce containers. Almost all the spices and herbs can be purchased in whole, ground, flaked, or powder for your convenience.

They also have dehydrated veggies for camping, soup mixtures, or special recipes. Celery stalk, pepper (green bell), and spinach flakes are among the choices.

SPICE MERCHANT   307-733-7811
P.O. Box 524
Jackson Hole, WY 83001

If you would like an Oriental market that makes house calls, contact **Spice Merchant** and let them be the "silent partner" in your cooking experiments.

Any ingredients that you could possible need for Chinese, Japanese or Thai/Indonesian cooking are now available in this very extensive catalog. Oyster, hot bean, Szechuan-style chili, and hoisin sauce are just some of the condiments that will change a simple stir-fry into a dining

experience. Roasted shrimp, lemongrass, and gado gado and pecel are a sampling of Thai/Indonesian products. All imaginable Japanese condiments and specialty fruits and vegetables are offered along with complete meals.

If you would like to experiment with these three unique cuisines, there are also very reasonably priced home kits containing the most often-used spices, vegetables, and supplies. Ready-to-heat and -serve foods along with flavorful teas and dried foods are part of their product line.

Spice Merchant presents a clear catalog that explains the products so the novice can cook like a native.

Cooking With Sugar Substitutes

*M*any people have dietary restrictions that do not allow them to have much, if any, refined sugar. Yet thousands of sugar-conscious people have discovered that substituting concentrated fruit juices, preserves, honey, and fruit products satisfy their sweet tooth in a delicious and naturally healthy way.

A general rule of thumb when replacing refined sugar in a recipe with one of the many products offered by companies featured in this and other chapters is: Substitute for sugar on a one-to-one basis and reduce other liquids by one third of the amount of sweetener called for in the original recipe.

With the "recipe" to eliminate refined sugar, you will eat healthier. In addition, you will find that supplementing the many available products will often improve your favorite dessert to four-star status.

VERMONT HARVEST   800-338-5354
Pan Handler Products   802-253-7138
Percy Hill Road
RR 4, Box 399
Stowe, VT 05672

Vermont Harvest's original conserve recipes are a taste sensation with a difference.

Use these delicious conserves on more than your morning toast. They are ideal as a filling for crepes, omelets, pastries, and as a substitute for sugar in baking. Some of your choices are pineapple-pepper jelly, banana-berry jam, and raspberry apple, to name a few. All would make a perfect hostess gift or thank you.

All of the conserves are blends of fruit, raisins, nuts, and spices, prepared naturally without additives or preservatives. Try some of the strawberry amaretto or apple rum in one of your favorite cake or muffin recipes. You will not be disappointed. Or with your toast, enjoy the blueberry bourbon or brandied peach to start the morning off right.

WAX ORCHARDS   800-634-6132
22744 Wax Orchards Road, S.W.   206-682-8251
Vashon Island, WA 98070

If your sweet tooth screams for dessert, but you have dietary restrictions on refined sugar, let **Wax Orchards** treat you to the sinfully delicious tastes you crave.

This unique company produces a large variety of naturally sweet and rich toppings, conserves, chutneys, condiments, syrups, and fancifuls. All are prepared without refined sugar. Fudge Sweet Toppings have received rave reviews from the *New York Times* and *Country Living,* and are offered in four flavors: Fudge Sweet original, amaretto, orange passion, and peppermint. They are perfect over ice cream, to add to warm milk for a satisfying drink, or in recipes such as brownies!

The All-Fruit Fancifuls are deeply fruit-flavored preserves that will liven up toast, muffins, ice cream, or cheesecake. You can make a sorbet, layered with slices of pound cake, or fill a tart shell with a Fanciful and top it with whipped cream for a quick, luxurious dessert. Some of the flavors include: raspberry, peach, orange-pineapple, strawberry, and blueberry.

WATKINS PRODUCTS   800-346-7976
Rt. 3, Box 113   214-377-3013
Frisco, TX 75034

For the highest-quality spices, herbs, syrups, and condiments, order from the oldest direct-mail food company in the United States, **Watkins Products** (established 1868).

Their catalog offers a wide range of everyday spices, herbs, and blends to complement even the best chef's kitchen. All are milled to retain the rich, natural oils with no salt added. The many blends offered enhance the taste of dips, sauces, salads, and mainstays such as meats, fish, or even an omelette or souffle.

They also offer pure and imitation extracts. The pure vanilla extract is a 34-percent mixture versus an 8-percent mixture, which you will commonly find in grocery stores. Among the many concentrates available: Smokehouse Barbecue Sauce, Barbecue Sauce, and Meat Magic. All can be used mixed or straight, but a small amount goes a long way. Juice and punch concentrates in flavors of black cherry, grape, lemon-lime, and others are perfect for parties or to set on the kitchen counter for individual glasses. Eight ounces makes sixty glasses at a cost of about seven cents per serving.

WOOD'S CIDER MILL   802-263-5547
RFD #2, Box 477
Springfield, VT 05156

If first comes the apple, and the only by-product you are aware of is apple cider, then contact **Wood's Cider Mill** for taste treats you will not regret.

For over one hundred years, this cider mill has been producing boiled cider and cider jelly. Boiled cider is a concentrate of apple cider about 7 to 1, cider jelly about 9 to 1. Nothing is added. The sweet, delicious flavor of 30 to 50 apples makes one pound of the cider jelly.

Use boiled cider with hot water for a soothing hot drink, in cooking as a glaze over ham, as a topping to pancakes, or over yogurt or ice cream. Use cider jelly with peanut butter for an award-winning PB&J sandwich, on muffins, toast, doughnuts, or with meat dishes.

Cider syrup, a special mixture of maple syrup and boiled cider, is perfect for French toast, waffles, or pancakes, or can be substituted in recipes instead of sugar in many cases. Maple syrup is naturally also available.

Try a Wood's Cider Mill sample pack, containing all four products, or send a gift of any combination.

Chapter 10

Health, Gift Packs and Other Items

Within the last decade, eating healthy has become more of a concern to many Americans. "You are what you eat," has finally sunk in. In the literal sense this is not exact, of course, or the strict vegetarian would turn into some cross between a carrot stick and a papaya. But cutting back on processed foods, restricting the intake of refined sugar and saturated fats, and balancing the diet to include smaller portions of meat protein and larger amounts of high carbohydrate foods is the healthiest approach.

This chapter contains an assortment of health-food companies along with articles concerning health. These businesses are mixed in with establishments that send gift baskets. All the companies in Feast will send gifts, but these firms specialize in the service.

BARTH-SPENCER   800-645-2328 Except NY
865 Merrick Avenue   800-553-0353 NY
Westbury, NY 11590

If you are looking for an extensive supply of vitamins and health aids, **Barth-Spencer** can supply you with the supplements you need.

They offer a large variety of natural formulas specially produced to assist and supplement your daily supply of vitamins and minerals. Products that help with lagging energy, nutrients that have particular qualities related to the eyes, and products that improve mental alertness are just some of the offerings. Others include dietary and cholesterol aids, and even tablets to curb your intake of sugar.

More standard vitamins and complex vitamins are also available. Most are conveniently packaged in tablet, softgel, or the all-day time-release form, captabs. Each supplement featured in their wide-ranging catalog also has information on the benefits of each product so that the best choice can be made for your eating habits.

Children of a Lesser Food

*H*ave you looked at the labels on food lately? Why are "flavor enhancers" added? What about the dyes, salt, sugar, and coloring, and are the vitamins added because the nutrients have been lost? Have toxins been processed in? If any of these queries concern you regarding an infant or small children, you may want to pursue the growing market of organic baby food and food companies.

A recent study by the Natural Resources Defense Council (NRDC) contradicts the parental ploy often used at mealtime: "Eat your vegetables, they will make you grow big and strong." The report, titled "Intolerable Risk: Pesticides in Our Children's Food," focuses on children's eating habits from birth to age five. In addition, the study outlines the quantities of pesticides used on twenty-seven of the most-often-consumed crops. The three most serious charges of the NRDC are:

● *The Environmental Protection Agency (EPA) does not take children's eating habits into account when setting legal limits on pesticide residues in food.*

● *The EPA does not adequately take into account the youngster's susceptibility to hazardous pesticides that possibly cause neurological and behavioral problems.*

● *The Food and Drug Administration's (FDA) sampling and testing procedures for illegal pesticide contamination are inadequate, and may even put the general public in danger.*

Youngsters consume more food for their body weight than adults. As a result, they receive four times the exposure to the eight carcinogens studied by the NRDC. In addition, an estimated 34 percent of the diet of preschoolers is fruit, compared to 20 percent for adults. The average preschooler also drinks eighteen times the

*amount of apple juice, compared to a grown woman. All these fac-
tors, and more, are the foundation of the NRDC study.*

*Diaminozide, sprayed on apples and other fruit to improve
shelf life and appearance, breaks down into UDMH (a potent car-
cinogen) in processed foods. Some examples are applesauce and apple
and other fruit juices. Organophosphate insecticide used on tomatoes, green beans, and other vegetables is possibly linked to nervous
system damage. The NRDC charges that children may be especially
vulnerable and possible victims of these carcinogens and toxins during the early years of their development.*

*In regulating pesticides on food, some say the government vir-
tually ignores the impact on children. Shannon Sullivan of the
NRDC comments, "There are a lot of scientists now that say children
are disproportionately impacted by chemicals when they are young."
Youngsters before five years old are especially vulnerable to carcinogens and neurotoxins, since their kidneys and neurological systems
are still maturing. Ms. Sullivan adds, "Children are at an
increased risk . . . their cells are dividing rapidly [compared to those
of older children or adults] and their immune system is not completely developed.*

*According to the NRDC study, 6,200 of today's preschoolers are
likely to develop cancer from the pesticide-tainted foods they consume
before age five! Fungicides which break down into ETU (which is
carcinogenic and toxic) are often used on tomatoes, potatoes, and
apples. Captan and Mancozeb are among these fungicides and may
lead to cancer, with a risk that is two to seven times the "allowable"
risk set by the EPA.*

*The conflict that the NRDC has with the FDA stems from the
low number of product samples tested: about 1 percent of production.
This routine testing can also only detect about one half of the pesticides that could be present. Additionally, the delays in laboratory procedures allow toxic foods to be stocked on the shelves of retailers and
end up on our dinner table.*

*Under mounting pressure from parents' organizations and
concerned health groups, the EPA banned the use of Diaminozide*

and the fungicide Captan on many crops as of winter 1990. This victory (resulting directly from the NRDC report) elated many officials and parents, but others felt the EPA was dragging its feet and should act more quickly to remove pesticides from the food supply.

The contention that eating food can be "hazardous to your health" bothers most of us, particularly when our children are at an even higher risk. Nutritionists certainly still recommend a balanced diet of fresh fruits and "lots" of vegetables. To stay away from heavily processed foods has always been wise and in the wake of the NRDC study seems even more important.

For more information or answers to questions concerning pesticides, herbicides, and fertilizers in foods, contact the National Pesticide Telecommunications Network at 1-800-858-PEST.

AMERICAN SPOON FOODS, INC.   800-222-5886
1668 Clarion Avenue   616-347-9030
Petoskey, Mich. 49770   Discount: Dried
                                  Cherries sample

For holiday or "just because" gift givers, **American Spoon Foods** offers a fabulous assortment of products to fit any taste or occasion.

Try some of their Spoon Fruits or fruit butters. The Spoon Fruits are preserves sweetened with fruit juice concentrates, no sugar added. Apricot, blueberry, and red raspberry are some of the ten offerings. Their fruit butters are puréed fruit (cherry, concord grape, and quince, to

---

name a few) that has a buttery consistency. They do not contain real butter and there is no sugar added.

Other unique edibles include pineapple-chili barbecue sauce, spicy ginger plum sauce, and peanut sauce. The Salad Dazzlers line includes herb-mustard, raspberry, and curry. About twenty preserves and jellies are features; native harvests, nuts, and regional tastes make up the balance of American Spoon Foods offerings. Gift baskets are available.

BALDUCCI'S   800-225-3822
Mail Order Division   718-786-4125 FAX
1102 Queens Plaza South
Long Island City, NY 11101

For over fifty years **Balducci's** has been serving New York with the finest foods from around the world. Today, you can experience this *giornalino* across America and sample rare delicacies gathered from small purveyors or find that special hard-to-find ingredient to enhance your favorite recipe.

Try the vegetable terrine, which is a melange of carrots, cauliflower, mushrooms, and asparagus baked in a seasoned béchamel sauce. To complement the mainstay—stuffed veal roast stuffed with fresh spinach, mushrooms, onions, and Parmesan cheese—include stuffed mushrooms or arancini, which is a combination of arborio rice with onions, prosciutto, Parmesan cheese, eggs, and chicken stock.

Perhaps pasta or seafood is on your menu. The gourmet selections, from lobster agnolotti to spinach tortelloni, or apple-smoked sturgeon to salmon mousse with pistachio sauce, are certain to entertain the palate of your guests with their true excellence.

CALLAWAY GARDENS   800-282-8181
Country Store
Pine Mountain, GA 31822

For wonderful country flavors of the South, put an order in to the famous **Callaway Gardens** country store.

Choose from the Georgia-cured ham, speckled heart grits, and muscadine delights. All are traditional food celebrating the best of Southern heritage. Explore the muscadine mystique with the Muscadine Six-Pack. The pleasures of muscadine grapes are confirmed by this pack, including two jars each of jelly, sauce, and preserves. Try a spoon of sauce over vanilla ice cream for a delicious treat.

Pick from the assortment of delicious preserves or jellies. The Preserve Six-Pack is an ideal gift containing red raspberry, blueberry, apricot, peach, and strawberry preserves. Or order a half dozen delectable fruit jellies, including blackberry, blueberry, crabapple, boysenberry, elderberry, and red raspberry. Each pack makes a perfect present or just something nice to send to say, "I'm thinking of you."

COOKIE EXPRESSIONS   800-443-5958
4227 North Buffalo Street   716-662-1071
Orchard Park, NY 14127

Cookie Expressions's delectable homemade cookies, bursting with rich chocolate chips, are a recipe for success.

Now you see it. Now you don't. A charming, tastefully arranged bouquet of chocolate chip cookies is a thoughtful and original gift for any occasion. The artful and attention-grabbing designs make them a delight to receive on birthdays, half birthdays (six months before or after the last, next year), baby showers, and for no reason at all.

Cookie bouquets are a delicious and delightful alternative to flowers. All are arranged in a wide variety of styles—with attractive tins, baskets, mugs, and more. If you need a special design, custom-made orders for the occasion are not a problem.

 CULPEPPERS POPCORN   800-426-0522 Ext. 7
14000 Dinard Avenue   213-921-9804 FAX
Sante Fe Springs, CA 90670

Everyone knows at least one popcorn fiend, and for that person, **Culpeppers Popcorn** can satisfy their every desire.

This company offers twenty scrumptious flavors of popcorn in decorative tins. The popcorn comes in 3½- and 6½-gallon reusable tins, with your choice of one flavor, or two-, three-, four-, or six-flavor combinations. The designs on the tins range from holiday scenes to nature prints, bears, ducks, and the classic tins.

A few of the flavors offered include regular air-popped, buttered, cheese, cinnamon, caramel, chocolate, peach brandy, root beer, and buttered rum. If you would like to try one of each of the flavors, order the Treasure Chest, which is loaded with twenty five-ounce bags of the most delightful popcorn flavors.

Dried Fruit

*D*ried fruits are simply fresh fruits with the water removed. Since the drying process does not adversely affect a fruit's nutritional value, most of its vitamins, minerals, and fiber remain intact. The main difference is that with the water removed, the bulk and weight of the fruit are greatly reduced, so its nutrients are concentrated in a smaller, lighter piece of food, making it easier to eat and transport.

On a fruit-to-fruit basis, dried and fresh are quite close nutritionally while ounce-for-ounce, dried fruits are by far a better source of potassium and iron as well as fiber. And dried fruit doesn't bruise! Enjoy!

EARTH'S BEST, INC.   802-388-7974
P.O. Box 887   802-388-9247 FAX
Middlebury, VT 05753

If your concerns about pesticides in your baby's food have heightened recently, you may want to consider the product line offered by **Earth's Best**.

This company grew out of the belief that babies deserve to eat foods that are free of pesticide residues which are possibly carcinogenic. They begin with pure, organically grown foods. The customized cooking process of the fruits and vegetables protects the nutrients that are often processed out.

Their product line consists of fifteen baby foods, juices, and cereals. Some of the vegetable choices include carrots, green beans and rice, and winter squash. The fruit servings range from apples and bananas, apples and plums to peaches, oatmeal and bananas, and plums, bananas and rice. Their juices are apple, apple-grape, and pear. They also offer brown rice and mixed whole grain cereal.

Earth's Best's complete line of products is not listed here and is continually expanding to offer more healthy options for your baby. (See also Children of a Lesser Food, page 199.)

Eat Your Way to Health

O ur nation goes to desperate extremes to stay healthy and thin. Our bloated-body society feels losing weight is the first step to a "better me"—hence, a diet is the answer.

Millions of overweight people have chosen from a plethora of diets to solve their "love handles" problem. Liquid formula, feast and famine on alternate days, low-calorie, even papaya and pineapple diets have been toted as fast slim-down methods—among hundreds of other programs!

Like most good results in life, success depends largely on consistency and a desire to succeed. Unless your menu is going to include a large dose of papayas forever, permanent weight loss is highly unlikely. An effective answer, supported by most physicians, is to change your diet, in other words, the types and quantity of food you eat. The second step is to develop a convenient exercise plan that does not have to involve a lot of sweating, expensive equipment, or health club memberships.

Good nutrition (a proper diet) starts with knowing your body's needs. Many sustenance items can be, and clearly should be, gradually cut back. A fourteen-ounce New York sirloin may taste great "going down," but supplies your body with far more protein than it needs. Instead of ordering or buying the fourteen-ounce, get a ten-ounce. Or one step better, buy the fourteen-ounce and split it with a friend, significant other, or even the neighborhood dogs! The point being, animal protein should be part of your healthy meal, but only a nominal portion, not the mainstay of the sitting.

Proteins available from plants are generally incomplete by themselves. This means two or more complementary plant proteins must be combined in the same meal to supply the body with the essential protein requirement. Just like meat, your body stores (in your "love handles") some extra fat and excretes the rest.

Ounce for ounce, fat in or on meat and the skin on chicken and fish contain twice the calories of most starch products. Potatoes slathered with butter, gravy, or sour cream become a caloric nightmare for those who are trying to trim down! Instead, try some fresh parsley, Parmesan cheese, and a touch of red wine vinegar on the potatoes. Trim the fat from steak or pork chops and remove the skin from fowl or fish. After a short time, you will smile when stepping on the scale.

Sweets are another culprit of the "over-an-inch pinch." Artificial sweeteners have never been shown to reduce weight or improve health. Part of the weaning process for the "sugarholic" is to have control. If your mission at the store is to buy a gallon of milk and you come back with milk, cookies, chips, and a couple of candy bars that looked lonely, you are out of control! The solution is either to bring only milk home, or to send somebody else for the milk if you lack willpower..

When baking, cut the ratio of sugar used in recipes by a quarter. Often, this won't destroy the taste or even be noticeable. Try one *teaspoon of sugar instead of two in the morning coffee. Once again, these are simple steps to a well-rounded diet which will lead to better health, a better self-image, and a higher energy level. The key to success is moderation and consistency.*

For Further Information:

Brody, Jane E. Jane Brody's Nutrition Book*: New York: Norton, 1981, and Bantam, 1982.*

Hausman, Patricia. Jack Spratt's Legacy: The Science and Politics of Fat and Cholesterol. *New York: Richard Marek, 1981.*

Reader's Digest. Eat Better, Live Better*: Pleasantville, N.Y.: Reader's Digest Association, 1982.*

FREEDA VITAMINS   212-685-4980
36 E. 41st Street
New York, NY 10017

Since the inception of the vitamin era, **Freeda** Pharmacy and Vitamin Company, founded in 1928, has been filling orders of superior quality vitamins to health-conscious people.

This company goes to extremes to ensure top quality and performance of each vitamin. Disintegration tests are designed for each product, to simulate the digestive system, that exceed established testing standards. They do not use coal-tar dyes, sulfates, sugar, artificial colors, or flavors. These products are also found on the Feingold-approved food listing for hyperactive children.

All the formulations are appropriate in purity for even the most orthodox vegetarian. Sources of binders and fillers, which are found in every vitamin tablet, are carefully selected and reviewed to guarantee purity, safety, and to ensure you receive "grade A" nutrients.

The dazzling selection offered by Freeda Vitamins is certain to contain the vitamins you take in various doses, quantities, and forms. Scan through their catalog soon, and for health-care professionals, take advantage of their courtesy discounts.

FIGIS, INC.   800-325-3690
3200 South Maple Street
Marshfield, WI 54449

For the last-minute gift pack in the holiday season or to please grumpy Uncle Ed on his birthday, consult the smorgasbord of choices offered by **Figis**.

Try a sausage and cheese box for the next party gift. Maybe a commemorative crock of sharp cheddar cheese spread will be perfect for a housewarming favor. Or if grumpy Unc has been agreeable for a change, send him the Standing Ovation, which includes over fifty samplings. He will have something to smile about with the six cheeses, a dozen cheese snacks, sausages, tortes, cakes, mustards, mixed nuts, and more.

Also throughout the catalog are clever occasion edibles. Slabs of cheese shaped in the form of a moose, fish, or cow and sausages that look like footballs (including the lacing) are only part of the gift-giving menu.

Sweets, nuts, and combination decorative containers with peanuts, pretzels, popcorn, and many other "munchies" are also only a phone call away.

HARMAN'S CHEESE & COUNTRY STORE   603-823-8000
Route 117
Sugar Hill, NH 03585

For a cornucopia of country gift items from New England, contact **Harman's Cheese & Country Store** for an unmatchable variety.

Cheese is certainly available for gifts or a reminder of New England. For tasty spreads, the cheddar in port wine with cognac or the variation with rum will be perfect for hors d'oeuvres. From the world's finest fishing waters, smoked salmon, crab meat, kippered fillets, and more can complement the cheese. Naturally, a wide variety of maple products, including syrup (of different grades), butter, granulated sugar, candy, and much more, are ideal gifts or essential items in your kitchen.

Hero and Dickinson preserves are offered in over fifteen flavors. Home-style Allberry Fruit Farm jam, made in New Hampshire, is available in wild blueberry, raspberry, raspberry-honey, peach-honey, and more. Many gift combinations, herb mixes, candies, and salad dressings from this country store can make gift giving easy and quite tasty.

HARRY AND DAVID   800-547-3033
Medford, OR 97501

For the very finest fruit and food gifts for family and friends, impress them with a selection from **Harry and David**.

Take advantage of the many combination gift baskets offered for that "thank you" to your housesitter or Aunt Ethel, who can't do enough for your children. The Willow Market Basket is delivered in a handwoven willow basket with antique green trim and inside Royal Riviera and D'Anjou Pears, crisp mountain apples, and a grand assortment of tasty treats: trail mix, fruit preserves, chocolates, and cheddar cheese.

Perhaps some baked goods would be more appropriate. A pumpkin cheesecake, or a nine-layer meringue-like chocolate torte are bound to end a meal with a special touch. Or select from the meat and seafood offerings for the mainstay of the meal.

Harry and David also has two food-of-the-month clubs. The first gives you a sampling of the many treats on a three-, six-, nine-, or twelve-month schedule. Some of the tasty treats included are Sweetheart's Cheesecake, Northwest smoked salmon, fresh peaches and pound cake, and chocolate truffles. The second club is offered on a three-, five-, eight-, and twelve-month schedule and delivers different fruits to enjoy throughout the year.

KNOTT'S BERRY FARM   800-877-6887
P.O. Box 1989
Placentia, CA 92670

For a cornucopia of delicious gift-giving packs, **Knott's Berry Farm** stands ready to ship.

One selection, the Gift From the Heart, is a tasty assortment of preserves, jams, and marmalades. The Box of Six has eight-ounce jars of apricot-pineapple, boysenberry, peach, and strawberry preserves, California plum jam, and orange marmalade. These presents come in boxes of four, six, nine, and twelve individual flavors to delight friends and family.

Perhaps the California Harvest of handpicked fruits, dried in the California sunshine, would be the right treat. This luscious selection includes sweet apricots, dates, figs, peaches, pears, and prunes carefully packed and delivered in a Knott's Berry Farm crate. Many baskets are available which incorporate taste treats like hot jalapeño jelly, cucumber and onion pickles, cornbread and biscuit mixes, and honey nut popcorn, ideal gifts for holidays or an exceptional thank you.

LAMBS FARM   800-52-LAMBS
P.O. Box 520   708-362-0742 FAX
Libertyville, IL 60048

For a fine assortment of country gift foods from the heart of our country, contact **Lambs Farm** for everything from cookies to cheese.

This company started thirty years ago as the dream of two teachers who believed that mentally retarded adults could and should have meaningful employment. Presently there are over ten facilities on the 63-acre site providing employment and services to the community. One is the country store, which features a wonderful assortment of gift food items.

Try one or two selections from the desserts, chocolates, or preserves varieties. The Lambs Own Rum Cake or the Lambs Amaretto Cake are perfect with a tin of their famous butter cookies for a festive dessert or afternoon tea. If your neighbor who "house-sat" would probably never wear an I LOVE LIBERTYVILLE T-shirt, the Lambs Preserve Wheel, containing six ten-ounce jars of preserves, is perfect. Flavors range from blueberry to orange-pineapple-cherry marmalade.

Also available are marvelous gift boxes, baskets, and even a wicker suitcase picnic basket filled with appetizing treats to bring a smile on any occasion.

LES TROIS PETITS COCHONS   212-219-1230
Pâtés & Quiches, Inc.   212-941-9726 FAX
453 Greenwich Street
New York, NY 10013

If pâté or terrines pass your lips once a year as a special-occasion nibble, the time has come to let **Les Trois Petits Cochons** introduce you to the delights of a weekly treat.

All the pâtés offered are made by hand with the finest, freshest ingredients...meats, vegetables, eggs, and cream...using spices, wines, and liqueurs as natural preservatives, never chemicals. Some of the traditional pâtés

---

offered are country pâté—coarsely ground pork with onion, garlic, and many herbs and spices—and green peppercorn pâté—finely ground pork with green peppercorns and cognac. Game pâtés use duck, rabbit, and venison as the base meat.

Many creamy, smooth, spreadable mousses are offered as the perfect crowd pleasers. The duck liver mousse, with pork and port wine, and truffle mousse—of chicken livers, pork, truffles, sherry, and Pineau des Charentes. Your choice of vegetable, salmon, and/or spinach, are also certain to please you or guests, whether served as heated sandwiches at lunch or as tasty delights during the cocktail hour.

NORM THOMPSON
P. O. Box 3999
Portland, OR 97208

800-547-1160
503-643-1973 FAX

Only Santa himself can top **Norm Thompson** in the overnight delivery business.

This company's catalog features a delicious array of edibles, from candies to salmon. Try some English candies for a last-minute gift. Or the perfect addition to any meal are the delightfully decadent Australian Dessert Apricots. These delicious plump, moist fruits are dipped in rich, dark chocolate to create an unmatchable taste treat.

During the holiday season, they publish two food-only catalogs. Within their pages you will find most any food item you can imagine—from old favorites to adorn your own festive table to perfect gift offerings—comprising the extensive repertoire that has made this company famous.

NUTREX HEALTH PRODUCTS   800-431-2582
634 Center Avenue
Mamaroneck, NY 10543

For an extensive selection of herbal juices and nutrient formulas to supplement your everyday diet, **Nutrex Health Products** will serve your needs.

This company's product line includes over thirty-five organically produced herbal juices, which aid in relieving almost fifty common ailments. The plant juices supply valuable nutrients and synergistic factors (the combination of many nutrients which effect is greater than the effect of the same nutrients taken individually) missing from all dried herbal products.

These products are 100

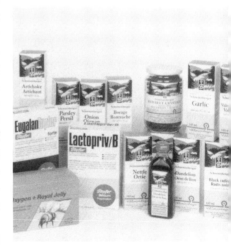

percent pure cell juices, not diluted, containing no additives whatsoever. This pureness provides an option for people who have adverse reactions to dried herbs due to fermentation or other factors. Whether the choice is artichoke or rosemary juice, each is clearly defined as to the nutrients they contain and in what cases they aid in improving health.

THE POPCORN FACTORY 800-858-1000
P. O. Box 4530 708-362-9680 FAX
Lake Bluff, IL 60044

What do guests, rented movies, and lunchtime have in common? Snacks, and that and more is what **The Popcorn Factory** can supply you with.

Send or receive the gigantic 6½-gallon silver can of popcorn with all butter or a three-way mix of butter, double cheddar cheese, and homemade caramel popcorn. Or enjoy Sandra Boynton's bears on the delightful 3½-gallon can

loaded with butter popcorn, mini-pretzels, and tortilla chips. For that muncher with a sweet tooth, send a plaid gallon pail of delicious chocolate chip cookies. So good, they're hard to stop eating.

Possibly, your recipient would prefer nuts and chocolates. They will be thrilled to receive the combination gifts with the assortment featuring regular, chocolate-covered, and honey-roasted peanuts; or with regular, chocolate-covered, and honey-roasted pecans. Many other combinations are offered.

S. E. RYKOFF & CO.   800-421-9873
P.O. Box 21467
Los Angeles, CA 90021

For an absolutely unmatched selection of gifts and unusual food items, **S. E. Rykoff & Co.** is at your service.

When you want to add a little color to your next party, have blue corn tortilla chips, blue popcorn, and pickled cactus—the tender green pads have long been a closely guarded secret in the Southwest. To add tantalizing treats to the main course, include—white asparagus and Belgian carrots.

But for a special gift that will not soon be forgotten, send the Fresh Caviar Sampler. This extravagant gift includes six one-ounce jars of beluga, sevruga, and osetra—all imported from the Caspian Sea—along with examples of America's finest roes—Golden, sturgeon, and salmon caviars. Many other size samplings are available.

To even attempt to list all the delicious ready-to-serve appetizers and cookies and cakes that S. E. Rykoff & Co. offers would be futile. You will need to look at their catalog.

SWEET ENERGY   802-655-4440
Box G
Essex Center, VT 05451

If you have a hard time keeping a steady supply of nutritious snacks on hand, **Sweet Energy** is ready to solve your problem.

After starting the company with mail-order dried apricots, their product line now consists of a smorgasbord of dried fruits, nuts, and nut and dried-fruit mixes, chocolate-covered nuts, fancy nuts, and more.

The Apricot Gift Sampler contains glazed Australian apricots, tart and tangy California apricot halves, Sweet Energy apricot granola, Turkish apricots...plus two apricot candies: apricot ambrosia and apricot coconut nuggets. After you've had your fill of these delights, try some of the specialty fruit, including nectarines, pineapple slices, cherries, apple rings, peaches, and more. All are dried and come in varying size packages, from eight ounces to five pounds.

Combination gift packs and party containers come with as assortment of dried fruits, mixed fruit and nuts, or chocolate-covered nuts and fruit. All are certain pleasers.

Sweet Energy also sends a monthly newsletter outlining specials and new products such as the Camper's Care Package for the lonely, homesick first-time camper. This is probably perfect for college students, too

---

SUMP'N SPECHUL   904-737-7600
6925 St. Augustine Road
Jacksonville, FL 32217

If you are looking for a gift, **Sump'n Spechul** can provide everything from cookies to caviar.

This company has some of the world's finest treats and gift baskets in their extensive catalog. A handsomely presented wicker gift basket filled with mouthwatering chocolates or a two-pound tin of Peber Nodder Danish cookies is a perfect "I love you" or thank you gift. Perhaps the Gourmet Delight, containing goodies delivered in a brass pail, will lift someone's spirits. You can choose from the "Sweet," a selection of dessert toppings and nut snack and miniature candy creations, or the "Sour," a selection of wine mustard, salad, or cookery or a combination of both.

Maybe a gift meal is in order. Choose from fettuccine pasta meals, a lobster feast, the sushi package, or the meat and fish offerings. Many other scrumptious gifts are available from Sump'n Spechul, so don't delay that overdue thank you or "I love you."

SUTTON PLACE GOURMET   800-346-8763
10323 Old Georgetown Road   301-468-7638
Bethesda, MD 20814

For a broad array of Europeon-style specialty gift baskets, **Sutton Place Gourmet** will help you celebrate your cherished occasion elegantly.

The Mediterranean Basket features the distinctive flavors of Greece, Italy, France, Turkey, and Spain. Among the many authentic Mediterranean products you will find inside are Lerida olive oil, Peloponnese roasted sweet peppers and eggplant salad, antipasto, balsamic vinegar, an assortment of special spices, and more. This gift is perfect for a lonely traveler, to remind them of the Old Country, or as a special gift for someone to try new tastes.

Do you have a college student who is "starving?" Send them the Student Care Package. The distant scholar will feast on salsa and dips, mixed nuts, a Toblerone bar, Baci hot sticks, Elliot's Amazing Fruit Juices, popcorn, and more. Who knows, they may even be inspired to write home.

THE HARRON'S OF SIMPSON & VAIL, INC.

P.O. Box 309   800-282-TEAS
Pleasantville, NY 10570   914-747-1336
                                                     914-741-6942 FAX

For more than 50 years, **Simpson & Vail** has been purveying the finest teas from around the world. To give your palate the taste it deserves, order some today.

Among the choice black teas are Assam, which is a zestful, invigorating North Indian tea noted for its mellow, malty flavor. Try Lapsang souchong from Formosa—rich bright

orange liquor with a distinctive smoky flavor. This is a treat for true tea lovers, but is an acquired taste. For further experimenting, green, oolong, Japanese green, and many other types furnish numerous temptations.

A wide collection of coffees as well as food items are a tradition. Fancy Vienna Roast or Indian Mysore may make the perfect gift. Confections, preserves, sauces, or salad dressings could be the perfect addition or elegant thank you to the housesitter or a colleague for a job well done. Spices and herbs and foods from around the world are only part of the almost endless selection of fancy edibles for your table or to be sent as gifts.

THE SWISS COLONY   608-324-5000
1112 7th Avenue   608-324-5050
Monroe, WI 53566

If you are looking for an overwhelming selection of appetizing cheese, meat, and sweets gift packs, **The Swiss Colony** can satisfy your order.

Each of the boxes offered includes a sampling of Wisconsin products. From single wheels of cheese to the colossal assortment of seventy-five items, including cheeses, sausages, tortes, petit fours, buttermints, cookies, nuts, preserves, mustards, and much more! Some gift boxes contain only cheese products, some just meat products, and many have both.

Fully cooked hams, turkeys, and ready-to-cook prime cuts of meat are available. Beef sausages in many flavors, steak strips, bratwurst, and even low-sodium beef products are just some of the meat offerings.

If the recipient of your gift has a sweet tooth, send one of the Dobosh Tortes varieties or luscious signature cakes. They are a guaranteed winner. The Swiss Colony also has food-of-the-month clubs.

WALNUT ACRES   800-433-3998
Penns Creek, PA 17862

When you purchase **Walnut Acres** foods, you can be certain they are pesticide-free and certified organic.

With a series of symbols next to the food items throughout the catalog, this company offers some of the "cleanest," pesticide-free products available by mail-order or through local food chains. The products are rated—Walnut Acres-certified organic: without synthetic chemicals since 1946; — Certified: producers comply with organic food certification: —Transitional organic: The producers are making a complete transition; no synthetic poisonous chemicals have been used on the product.

Within the extensive catalog you will find a plethora of products. You would have organic coffee for breakfast, one of the hearty hot cereals, a slice of bread, and juice. Lunch could consist of soup, a sandwich with their Famous Peanut Butter, or one of the farm-fresh cheeses and a piece of naturally ripened fruit. Dinner could include a healthful pasta and sauce dish with vegetables, organic meat dishes, and one of their luscious desserts.

WILLIAM-SONOMA   800-541-2233
P.O. Box 7456   415-421-5153 FAX
San Francisco, CA 94120

Last-minute gift shopping for special foods does not have to be hectic and hopeless if you have a **William-Sonoma** catalog.

Choose from the dozens of offerings, from fancy Valrhoma Carre de Guanaja after-dinner chocolates to a wheel of stilton cheese—blue-veined, sharp, and aged. Or perhaps cooked Virginia peanuts or a bag of pecans will satisfy that almost-missed birthday or needed thank-you. Along with the food items you could also send utensils such as nut-crackers.

When the holiday season approaches, enjoy sending cakes, an Italian Panforte, or a box of elegant chocolate truf-fles, treats the recipients would not ordinarily buy for them-selves. And if you order William-Sonoma Coconut and Caramel Gold Bars, be advised to keep them in a bank vault—they disappear mysteriously.

ZABAR'S   212-496-1234
2245 Broadway
New York, NY 10024

At Zabar's, they roast and blend over 30,000 pounds of cof-fee beans a week, but that is only a part of their extensive, delicious food offerings.

Feel free to order your special coffee blend or some croissants and caviar. If you are food gift shopping, their cata-log will offer hundreds of choices from the exotic to more typical items. So if you really need game rabbit pâté, just con-sult the catalog. But if you need a gift for Aunt Gertrude's birthday, and eating pizza is radical to her, choose from the fine Droste chocolates selection.

Try the two Auricchio cheeses that won the title "World's Best" at the World Natural Cheese Championship: the Americano Provolone and Parmesan. Or choose from the hundreds of imported and domestic cheeses in their renowned cheese department.

Subject Index

Baby food, 205–6
Baking ingredients
  assorted, 62–63
  flour, 50–51, 58–59
  general information about,
    55–56
Breads
  bagels, 59
  croissant, 52
  English muffins, 70–71
  flour for, 50–51, 58–59
  mixes for, 51, 58, 60–61, 66
Brownies, 17–18, 69

Cakes
  assorted, 53–55, 60, 63–64,
    69–70, 211, 213, 222
  cheesecake, 57–58, 61–62,
    68–69
  fruitcake, 29, 39, 51, 52–53,
    56–57
  Mardi Gras King Cake, 59–60
  rum cake, 65
Candies, 17, 23–24, 30, 214, 219,
  222, 223–24. See also
  Chocolates

Cheese
  assorted, 35, 36–37, 40–43,
    45, 209–10, 222, 223, 224
  cheddar, 34–35, 44, 81
  general information about,
    33, 37–38, 43–44
  goat, 36
  Swiss, 44–45
  Trappist, 39
Chocolates. *See also* Candies
  assorted, 19, 27, 28–29
  general information about,
    20–21, 27–28
  shells, 16
  Swiss, 18
Coffee, 35, 50, 163–64, 165,
  221, 224
  general information about,
    161, 164–65
Condiments
  assorted, 181–82, 202
  barbecue sauce, 76–77, 182,
    184, 194
  garlic, 137–38
  general information about,
    175–76

herbs and spices, 181, 185,
186–88, 190–91, 194
honey, 64, 176, 178–80
maple syrup, 35, 42, 80
mayonnaise, 89
mustard, 89, 186
preserves, 74, 121, 131, 177,
180, 192, 193, 195, 201–2,
203, 210,212, 213
relish, 180
sausage seasonings, 183–84
vinegar, 182–183
Conserves. *See* Condiments,
preserves
Cookies, 17–18, 48, 51, 64–65,
66–67, 69, 77, 204, 213,
217, 219, 222
Curd, 189

Entrees, prepared, 148, 155,
202, 219, 223

Flour, 50–51, 58–59
Flowers, use of in garnishing,
188
Fruits
apples, 88, 124, 127, 133
assorted, 122, 137, 211
dates, 130, 139
dried, 17, 127, 128, 130–31,
139, 140–41, 212, 218
dried, general information
about, 205
general information about,
119–20, 122–23
grapefruit, 127, 134, 140
oranges, 121, 133, 134–35, 140
organically grown, 125–26
pears, 124, 127, 133, 137
tomatos, 141

Ice cream, 48–49

Jams. *See* Condiments, preserves
Jellies. *See* Condiments, preserves
Juices
concentrates for, 194
apple cider, 195
herbal, 215–16

Liquor, 77, 167, 174. *See also*
Wine

Mail-order industry, the
consumer protection agen-
cies, 6–8
food-by-mail clubs, 6–68
general information about,
2–4
name removal services, 8–9
tips for shopping through,
4–6
Marmalades. *See* Condiments,
preserves
Meats
assorted, 149, 158–59, 222
bacon, 80
barbecued, 150–51, 153,
159–60
beef, 147, 148, 154–55, 158,
160
bologna, 149
chicken entrees, 148, 155
game, 74–75, 152–53
ham, 144–45, 151, 160
pork, 146–47, 154
sauces for, 94–95
sausage, 44–45, 80–83, 144,
145, 147–48, 150, 152
seasonings for, 183–84

Vendor Index

Alaska Sausage & Seafood Co., 74
Alaska Wild Berry Products, 74–75
Alaskan Gourmet Seafoods, 104
Amber and Company, 48
American Spoon Foods, Inc., 201–2
Astor Chocolate Corp., 16
Atlantic Seafood Direct, 104–5

Bacino's, 75–76
Balducci's, 202
Bandon Foods, Inc., 34
Barth-Spencer, 198
Bates Bros. Nut Farm, 17
Ben & Jerry's, 48–49
Best of Kansas, 76–77
Bette's Diner Products, 49–50
Bewley Irish Imports, 77
Bland Farms, 120–121

Blue Heron, 121–22
Boomer's Oogies, 17–18
Brown & Jenkins Trading Co., 163
Brumwell's Flour Mill, 50–51
Buckmaster, 144
Burger's Ozark Country Cured Hams, 144–45
Burrito Express, 78
Byron Plantation, 78–79

C'est Croissant, Inc., 52
Cabela's, 145–46
Cafe Beaujolais, 51
Calef's Country Store, 34–35
Callaway Gardens, 203
Caviarteria, Inc., 105–6
Chapita's, Inc., 79–80
Cheese Junction, 35–36
Children's Catalogue, 123–24
Chops of Iowa, Inc., 146–47